Leading in the Digital World

Management on the Cutting Edge Series from MIT Sloan Management Review

Edited by Paul Michelman

Published in cooperation with *MIT Sloan Management Review*

The AI Advantage: How to Put the Artificial Intelligence Revolution to Work
Thomas H. Davenport

The Technology Fallacy: How People Are the Real Key to Digital Transformation
Gerald C. Kane, Anh Nguyen Phillips, Jonathan Copulsky, and Garth Andrus

Designed for Digital: How to Architect Your Business for Sustained Success
Jeanne W. Ross, Cynthia M. Beath, and Martin Mocker

See Sooner, Act Faster: How Vigilant Leaders Thrive in an Era of Digital Turbulence
George S. Day and Paul J. H. Schoemaker

Leading in the Digital World: How to Foster Creativity, Collaboration, and Inclusivity
Amit S. Mukherjee

Leading in the
Digital World

How to Foster Creativity, Collaboration,
and Inclusivity

Amit S. Mukherjee

The MIT Press
Cambridge, Massachusetts
London, England

This book was set in Stone Serif and Stone Sans by Westchester Publishing Services. Printed and bound in the United States of America.

Library of Congress Cataloging-in-Publication Data

Names: Mukherjee, A. S. (Amit Shankar), author.
Title: Leading in the digital world : how to foster creativity, collaboration, and inclusivity / Amit S. Mukherjee.
Description: Cambridge, Massachusetts : The MIT Press, [2019] | Series: Management on the cutting edge | Includes bibliographical references and index.
Identifiers: LCCN 2019031915 | ISBN 9780262043946 (hardcover)
Subjects: LCSH: Technological innovations. | Creative ability. | Leadership.
Classification: LCC HD45 .M845 2019 | DDC 658.4/092--dc23
LC record available at https://lccn.loc.gov/2019031915

10   9   8   7   6   5   4   3   2   1

For Eric Mukherjee
Scholar-in-Training

and

Vijay Ghei
Extraordinary Creator

Contents

Series Foreword   ix

Preface: What Can Yet Another Book on Leadership Possibly Say That Is Different?   xi

Acknowledgments   xv

1   The Births and Obsolescence of Leadership Ideologies   1

I   What Do Digital Technologies Really Do?   15

2   Restructuring Work and Organizations—Six Principles   17
3   A Digital, VUCA World—The Seventh Principle   33
4   Implications for Leadership   41

II   Leading for Creativity   57

5   The Many Faces of Talent   59
6   A Broad Wingspan, Not a Long Tail   75
7   Being Truly Collaborative   85
8   Championing Creativity   99

III   Guide Rails for Creative Efforts   121

9   Defining the Uncrossable Lines   123
10   Developing Your Strategic Intent   137

IV   Where Next?   159

11   Building a Personal Leadership Philosophy   161

Notes   167
Index   199

Series Foreword

The world does not lack for management ideas. Thousands of researchers, practitioners, and other experts produce tens of thousands of articles, books, papers, posts, and podcasts each year. But only a scant few promise to truly move the needle on practice, and fewer still dare to reach into the future of what management will become. It is this rare breed of idea— meaningful to practice, grounded in evidence, and *built for the future*— that we seek to present in this series.

Paul Michelman

Editor in chief
MIT Sloan Management Review

Preface: What Can Yet Another Book on Leadership Possibly Say That Is Different?

Amazon.com recently offered over 50,000 results when I searched for "business leadership books." So, you may justifiably ask, "What makes this book different?" *Don't* buy, borrow, or read this book if my response doesn't (at least) intrigue you.

Leadership Books Typically Adopt One of Two Fallacious Perspectives

The first fallacious perspective is "Emulate this great man (usually a man) who behaved in this manner/acted thus/had these characteristics." The man is usually a very extroverted and very rich American or British of Caucasian ethnicity. These characteristics collectively exclude most people worldwide—including most Americans, Brits, Canadians, Australians, and New Zealanders. Yet, they bias our views: Even recently, I've heard superstar leadership professors at top schools extol the virtues of a square jaw and advise women to lower the pitch of their voices. Even if you personally conform to this flawed stereotype, you can't afford to believe it. Quite often in the digital world, the people who'll crucially influence (your) success won't conform to it.

The second fallacious perspective is "Our research (with tons of data) shows that these competencies are key for leadership." These "competency models" unavoidably assume that competencies that were useful in the past continue to be so. Moreover, they often embody unfounded cultural assumptions. For example, one suggests that "dutiful" people are less likely to protect subordinates from superiors. This devalues the performance of billions of people from cultures where not being dutiful is explicitly dishonorable. No wonder that many Pacific Asian business executives want "Asian" models of leadership. But that's a flawed solution: Instead of substituting one cultural paradigm for another, business leadership standards in the digital world must embrace diverse peoples equally.

Moreover, Leadership Models Typically Ignore the Oversized Role of Technology

Leadership experts present their percepts as ageless and universal, rarely mentioning technology. Digital technology experts see leadership merely as a tool necessary for guiding complex projects. Both groups ignore the changes digital technologies are demanding of leaders.

A brilliant academic study of centuries of coevolution of technologies and business identified "epochal" transitions that produced tumultuous change. Each time, sets of novel technologies—like today's digital technologies—raised corporate performance by orders of magnitude. They had long arcs of impact—decades, even centuries, later, we still use them. Not surprisingly, embracers often thrived; those that didn't inevitably faltered.

Embracing these technology sets wasn't easy: The long-arc-of-impact technologies demanded very difficult transformations of work, values, and organizational structure that changed companies and even transformed societies.

Leading in the Digital World differs from all other leadership books because it provides a model of leadership that reflects the impact of digital technologies. Instead of jumping onto the bandwagon of "disruption"—a grossly overused term that has gotten disconnected from its originally intended meaning—it describes precisely how digital technologies are upending the nature of work and organizational structures and, hence, the context for leadership. *At their core, these changes enable, even require, leaders to engender creativity—the ability to look past received wisdom and traditional approaches to give form or structure to new ideas.* This makes the digital epoch profoundly different from the prior two epochs of twentieth century; in those, business leaders relentlessly drove productivity.

Regardless of your gender, nationality, or culture, you need to adapt. If not, like your predecessors who resisted prior long-arc-of-impact technologies, you may be left behind.

My Expertise in These Issues Goes Beyond the Academic Research That Underpins the Book

The Leader in a Digital World draws on research in multiple academic disciplines and on research I specifically did for this book. With generous funding from IMD, where until recently I was Professor of Leadership and Strategy, I conducted a 700-repondent survey of global executives and interviewed business leaders around the world. Additionally, this book draws on my own prior research published in peer-reviewed journals, a book, practitioner-oriented media (including *Sloan Management Review, Harvard Business Review*, and *Forbes*), and work-for-hire reports for leaders of global companies.

Equally important, the book draws on lessons I learned during the (almost) two decades I worked outside academia. As a (co-)practice leader in two global technology consultancies and as the chief technology and strategy officer of a small NASDAQ-listed company, I spearheaded "can't be done" digital and nondigital technology projects. While I was at those consulting companies and later, at my own boutique consultancy, I also advised CEO/CXO-level executives of well-known global companies on leadership and strategy challenges that digital technologies pose.

Regardless of whether you're a current executive or a business student, this book will prepare you to lead in environments that prize developing and executing new ideas over making existing ideas faster or cheaper. Instead of decreeing lists of "five things to implement immediately," it will address mindsets, behaviors, and actions you should embrace, explaining why these matter. Personal change takes time, so it will also help you prioritize issues relevant to you right now.

Acknowledgments

This book has been in active development for three years and informally percolating in my mind for at least a couple more. During this time, I talked to executives and managers and interviewed many formally. Though only a small handful of those interviewed are named in these pages, I am grateful to all of them for sharing with me their time and experience-based wisdom about leadership in their companies and/or their parts of the world.

IMD's faculty research budget funded a good part of the research. This work has undoubtedly benefited from that support.

Five people helped me immensely. Dr. Amitabha Chaudhuri, the chief technology officer of a biopharma company, spent hours explaining to me how the drug discovery process has changed as a result of the sequencing of the human genome. Since he kept cautioning me not to simplify complicated issues too much, any errors that remain are mine alone. My dear friends, retired engineer Vijay Ghei and IMD professor Michael Watkins, served as sounding boards for all my early ideas. They delved deep into their unique bases of experience to improve several of them. The "Principles" in particular benefited from their doing so. Another friend, Todd Rhodes, read the first draft of every chapter and gave feedback. He doesn't believe it, but his observations helped a lot. Emily Taber, an acquisitions editor at the MIT Press, not only agreed to publish the book but also pushed hard for changes that have made it much stronger. Until a few weeks ago, I had no idea why authors thanked their editors; now I know.

Academic research works because writings are blind peer-reviewed. It is a thankless task, and I applaud those who volunteer to do so, including the four who opined on my ideas.

1   The Births and Obsolescence
of Leadership Ideologies

Those who cannot remember the past are condemned to repeat it.
—George Santayana, *Reason in Common Sense*, Vol. 1

Let's begin by fleshing out and building on the technology-based perspective of leadership. Throughout history, people have excelled at killing each other in bloody wars. All over the world, museums house the
artifacts they used. Governments have always defined "national security" as their prime responsibility and spent uncountable amounts on
weapons, sometimes at the expense of other key needs; first courses in
economics often present this as the "guns versus butter" challenge. Thus,
throughout history, weapons makers have been at the forefront of technological and organizational change. What they did trickled out into
the rest of the economy, first to other manufacturers and then to others.

In the mid-1980s, former Harvard Business School professor Ramchandran Jaikumar got to study almost five hundred years of gun making at one company, Beretta.[1] Through all of Europe's blood-drenched
history, the same family controlled it. Their detailed records gave him
the opportunity to study the coevolution of business and technology in
the context of one cutting-edge industry. Since the basic structure of a
gun didn't change over time, he could isolate what did change.

Jaikumar described multiple "epochal" transitions that produced
tumultuous change in businesses and society. Each time, a set of novel
technologies raised quality and productivity by orders of magnitude.
Each set had a long arc of impact; decades, even centuries, later, we still
use it. Companies had to embrace them merely to survive.

Embracing wasn't easy: The long-arc-of-impact technologies demanded very difficult transformations of work, organizational structures, numbers of employees, knowledge needs, what work was done and how, what skills were key, and organizational values. Effectively, every few decades the meaning of "company" changed profoundly.

Around 1800 CE, the micrometer and third-angle-projection engineering drawings engendered the "English system." These gave workers the ability to control what they worked on and check their own work. Companies began hiring workers directly and grew in size, while guilds collapsed, ending indentured servitude. A few decades later, go-no-go gauges and specialized machines powered the "American system." Whereas the English system required *precision* that only *skilled* workers could master, the American system emphasized the opposite: the *maximum acceptable clearance* between mating pieces. This seemingly minor change enabled mass manufacturing: Companies employed many more *unskilled workers* and even outsourced work. Socially, masses of people moved from farms to cities for factory jobs.

Notably, we still use the long-arc-of-impact technologies that powered those epochs: micrometers, third-angle-projection engineering drawings, and go-no-go gauges. The roles these forgotten technologies played were at least on a par with other better known ones—such as the steam engine—that enabled industrialization.

Scientific Management and the Age of Authoritarian Leadership

At the turn of the twentieth century all the conditions necessary for mass production in large factories converged. Interchangeable parts reduced the precision—and hence, the time and effort—needed to create each workpiece. Electric power allowed many machines at each site. Edison's light bulb, the end product of about a century of innovation by multiple inventors, enabled many people to work away from natural light.

And so, Frederick Winslow Taylor got the opportunity to increase people's productivity. Armed with stopwatches and clipboards, Taylor studied the minutiae of work—how people moved, what they did and why, and how much time each element of work required. He discovered that efficiency dropped when people got tired or hungry or needed toilet breaks. Those who did many things weren't equally proficient at all. Moreover, even when assigned carefully delineated tasks, some people discovered how to outperform their compatriots. Taylor concluded productivity could be improved if workers focused on narrow tasks, followed prescribed processes, and got appropriate rest breaks. Over a century later, industrial engineers still use similar "time-and-motion studies" to design work, including for robots.

These ideas led to the formation of functions, narrowing of tasks assigned to individuals and the elimination of discretionary effort. By instilling the "one best way" logic, they shifted power inexorably to managers, who could now prescribe what each worker had to do, when and how, without explaining why. They could also punish failure to conform to standards.

Workers could no longer see the logic of the entire process. Assembly lines—designed using time-and-motion studies—became a key system of production and represented the complete antithesis of human collaboration. Many factories forbade communications even among people who worked on them side-by-side.[2]

More broadly, as companies replicated the all-knowing, efficiency-focused, number-crunching, "one best way" logic among managers at all levels, they became giant machines and people became mere cogs within them. De-skilled, people worked in functional departments with rigid boundaries that we today decry as "silos."

A person's manager always knew best, even when he didn't. It didn't take much for managers to emulate the hard-charging, take-no-prisoners, brook-no-opposition style of the era's robber baron entrepreneurs. The Authoritarian Leader had arrived in the business world. He ruled the roost for decades: in Japan till the 1950s/early 1960s, and *in the West till the mid-to-late 1980s*.

Authoritarianism often—though not universally—took a damaging form that went far beyond expectations of unhesitating obedience. *Fortune*, the premier business magazine, periodically published articles entitled "America's Toughest Bosses." The 1989 article[3] discussed seven out of more than fifty finalists. "Toughness," it noted, could be for "constructive and legitimate purposes" but also "pushes people to their limits or beyond for capricious or inconsequential purposes." These business leaders were all verbally abusive; some even threw heavy objects at subordinates who displeased them. So *Fortune* found it necessary to add, "*We uncovered no violence*" (emphasis added).

The last "America's Toughest Bosses" article—published in 1993![4]—began with the words, "In an era of endless restructuring, cutting heads like Robespierre on a rampage is just average. These leaders inflict pain by messing with your mind as well." After excluding those who had been on prior lists and "thoroughly" investigating the finalists—more than 70 "of corporate America's hardest-nosed head persons"—it too featured seven CEOs. The journalists observed that

> over-the-top toughness today implies something more—something the taskmasters who made our final cut possess in spades. Think of it as a penchant for psychological oppression—an especially sadistic way of making a point, say, or a bullying quality that can transform underlings into quivering masses of Jell-O. Says Abraham Zaleznik, a professor emeritus of leadership at the Harvard business school: "Tough is passé. Today you're dealing with a variety of head games. That's where the cruelty is." Sometimes driven demandingness can be downright abusive.

The lists—and the finalists—included many of the best known American CEOs and renowned S&P 500 and Fortune 500 companies. Consider just a few: Edwin Artzt (Procter & Gamble), Robert Crandall (American Airlines), Maurice Greenberg (American International Group), Andy Grove (Intel), Steve Jobs (then at NeXT), Andrall Pearson (PepsiCo), Donald Rumsfeld (Searle) and Jack Welch (GE). The monikers used by their compatriots, or assigned by *Fortune*, are also telling: Prince of Darkness, Jack the Ripper, The Pompadoured Bully, Dr. Jekyll and Mr. Hyde—a "thoroughgoing SOB—cold, calculating, and mean." Clearly, authoritarianism wasn't an aberration; it is how American companies were regularly led.

In a 1986 *New York Times* article,[5] Dr. Daniel Goleman, now considered one of the most influential researchers on leadership, noted, "Good bosses, experts say, are a more uniform lot: Typically, they know the company's business and perform their assigned tasks. At the same time they help employees grow, give credit where it is due, dole out criticism where it is needed, and create an atmosphere in which it is easy to talk." Note that he didn't write about self-starting, or inspiring, or out-of-the-box thinking, or any of the other characteristics we look for today.

Goleman presented striking contrasts to this "good boss." He called Juan Trippe, the Pan American Airways chairman who was one of the best regarded leaders in the 1960s, "mean-spirited" and "capricious." He described Charles Revson, who headed the cosmetics company Revlon, as "a ruthless, crude, arbitrary whip-cracker" under whom "paranoia prevailed." Quoting a recently published book, he said that Henry Ford II used to "say to himself in the morning as he shaved, 'I am the king, and the king can do no wrong.'"

He also reported on a Center for Creative Leadership study of executives from eight major corporations. As you read this extract, keep in mind that those opining were among the winners in this Darwinian world, not average midtier executives or junior employees:

> As part of the study, highly successful executives, all among the top 100 managers in their corporations, were interviewed about key events in their careers and the roles bosses had played.
>
> The most frequently mentioned villain was the "snake-in-the-grass" boss, one who lacks basic integrity. ... Also high up on the list: the Attila-the-Hun boss, who bulldozes his way over any objections and takes offense if others make their own decisions or stand out in any way. ... Next in frequency was the heel-grinder boss, one with no respect whatsoever for employees. Such bosses belittle, demean and humiliate those beneath them, and will, for instance, fiercely criticize an employee in front of a group.

Not surprisingly, in the early 1980s, US business schools were still teaching about this take-no-prisoners style of leading. For example, at the brand-name school where I did my MBA, we discussed a case study about Harold Geneen, the chairman and CEO of what was then one of America's very large, admired companies, ITT.[6] A video that accompanied the case showed him belittling one of his executive officers for not delivering a required result. Half of my class found his doing so to be acceptable. The other half was uneasy, not quite sure that publicly castigating a senior officer was the right way to get results. No one, including the professor, said it was flat-out wrong. Nor did anyone question the morality of releasing a video that had the potential to show a person's humiliation to business students worldwide. No ethical concern had stopped the Harvard professor who had created the case or others involved in its production. It was after all, business as usual in America.

The two *Fortune* articles weren't exceptions; they are the only ones available online. A 1995 article in *Psychology Today* [7] noted, "[T]here's the issue of *Fortune* magazine devoted every couple of years to America's 'toughest' bosses." It summarized the thoughts of Professor Harry Levinson, a leadership expert on the faculty of the Harvard Business School, on bullying in organizations:

> 40 years of consulting have given him some idea of what they do and why. They over-control, micromanage, and display contempt for others, usually by repeated verbal abuse and sheer exploitation. They constantly put others down with snide remarks or harsh, repetitive, and unfair criticism. They don't just differ with you, they differ with you contemptuously; they question your adequacy and your commitment. They humiliate you in front of others.

Quoting another consultant, the article noted that "[a]t least in large organizations, bullying is not as blatant as it once was. 'The John Wayne image of leader doesn't go over so well in the '90s. ...' Intimidation tends to be more polished." However, it added that bullies still "are positively thriving at small companies."

Leadership ideas we recognize today—and wrongly consider timeless—actually originated in Japan in the middle of the twentieth century.

The Quality Movement and the Birth of Empowerment Leadership

The next epochal change, which brought with it empowerment of people, was decades in the making. AT&T's renowned Bell Labs invented the core technology that sparked it—statistical process control—in the 1920s. Though armories adopted it in the 1930s, its use didn't spread. Ignored in America, it brought salvation to a Second World War–devastated Japan.

In 1950, W. Edwards Deming introduced statistical process control to Japan. His teachings negated the essence of Taylorism—measuring an individual's efforts—and emphasized collaborative groups and work systems. A few years later, Joseph Juran taught Japan that "quality management" must start from the top. Largely unrecognized at home, these two Americans became revered in Japan, which named its two highest national awards for quality after them. Japan's homegrown experts—including Kaoru Ishikawa, Genichi Taguchi, Shigeo Shingo, and Taiichi Ohno—greatly expanded the epoch's technology set.

Japanese companies discovered that quality technologies required collaboration among groups of workers. They created Quality Control Circles, which were later called Quality Circles, and then, very simply, teams, or if appropriate, cross-functional teams. These "empowered" workers to make key decisions after collective, data-based deliberations. Being on the front lines, the workers knew what needed improving. Their managers functioned as team leaders and coaches, clearing roadblocks so that the teams could function effectively. Outside Japan, managers typically made the types of decisions groups of Japanese workers ably made.

Canadian novelist Arthur Hailey wrote several books that described the inner functioning of different US industries. Read *Wheels*[8] (1971), about the automobile industry, and two things will strike you: First, the Taylorism-driven industry routinely produced good and downright shoddy cars on the same assembly lines. Second, at dealerships, Americans could custom order the precise cars they wanted and get them in about three months. Today, these facts are mostly forgotten.

By 1981, Japanese companies were seriously challenging American companies. Using the full power of the quality movement epoch, they attacked both these realities. Their cars weren't showy, but they were problem-free. Customers couldn't get the precise cars they wanted but could choose from available "option packages" and drive off with new cars immediately. The US auto industry was under attack not just in the quality of its products but also in its business model.

Japan's indisputable edge in quality challenged many other industries. David Garvin published an influential study with incontrovertible data that showed that the worst Japanese manufacturer in a major industry had a much lower product failure rate than the best American manufacturer.[9]

After trying to adopt the simplest tools of quality without making any organizational changes, in the mid-1980s, belatedly, America began learning an important lesson: It couldn't embrace quality technologies without making organizational changes. Consistent with Taylorism, its companies had very steep hierarchies in place. These had to go.

Kim Clark and his doctoral student (now professor) Takahiro Fujimoto studied product development in 20 automobile companies in the United States, Europe, and Japan. In the late 1980s, these accounted for most of the global production.[10] Since this industry generally set the benchmark for many others, it revealed how work—in particular, knowledge work— was done in these regions. Scholars of technology and manufacturing, not leadership, they used terms like "specialization," "internal integration" (linking technical areas), and "external integration" (responsibility for interpreting market requirements).

Clark and Fujimoto wrote, "A very narrow specialist might have responsibility for initial design of a small part, such as a *left* rear tail light" (emphasis added; the design of the *right* light was presumably someone else's responsibility). Elsewhere, they observed, "Many of the low performers, mostly US and European firms, are so highly specialized" that it "creates problems of coordination. Nor does the high specialization...necessarily translate into superior product performance." Their specialization index for US companies was 127% higher than for Japanese companies; Europeans fell in the middle, 56% to 65% higher. That said, taking into account other evidence, they opined, "Whether one looks at volume producers or high-end specialists, European projects were more functionally oriented than the US or Japanese products."

Their data clearly showed that while Japanese firms had adopted cross-functional integration with team leaders by the mid-to-late 1970s, "[s]ome of the US and European car producers...in recent years conducted extensive in-house studies of the organizational patterns of seemingly effective Japanese competitors." Several started making serious changes in the mid-1980s. The professors opined that among very highly specialized, high-end European car manufacturers, "Strategic-organizational changes...are likely to be seen in the early 1990s."

As implied by the 1993 *Fortune* article ("era of endless restructuring, cutting heads like Robespierre on a rampage"), dismantling hierarchies and adopting flatter, team-driven structures took a devastating toll. If

I asked you to guess the last time unemployment and corporate bankruptcies reached the levels they did during 2008–2009, you, like every executive I've asked this question, would probably have said, "The Great Depression of the 1930s." US government data show that the real answer is the mid-1980s. During those tumultuous years, many among those who lost their jobs were middle managers for whom Harold Geneen's behavior had been an everyday fact of life.

Social and organizational changes don't happen instantaneously and in their full-fledged forms. Jaikumar's research found that early adopters of epochal changes did so within fifteen years of the emergence of the technology, but broad adoptions took up to fifty years. So epochs overlap at their beginnings and ends—as the scientific management and the quality movement epochs did in the 1980s and even in the early 1990s.

The long periods of epochal change also explain why teams occasionally appeared sooner than in this discussion. Lockheed had its famous Skunk Works since 1943. In the early 1960s, the IBM 360 was famously created by a team that practiced flat, not hierarchical, communications among peers. At its Kalmar plant, Volvo adopted collaborative manufacturing practices as early as the mid-1970s. But these exceptions, usually in flagship high-technology projects, don't describe how work was normally done.

My brand-name business school had partially anticipated changes. Along with Harold Geneen and Taylorism, we discussed the best-selling book *The Art of Japanese Management*[11] in our strategy classes, and we debated the topic repeatedly in our operations and our organization behavior classes. We studied Japanese culture and learned about community centricity as a complement to American individualism. Perhaps reflecting the impending demise of authoritarianism, our class studied a module on ethics, supposedly the first ever in any premier business school. Yet, many of my classmates and some professors weren't convinced. America was too individualistic a society, they proclaimed, to adopt fads like teams.

Beginning in the early 1990s, books on teams and leaders who were the antitheses of their authoritarian predecessors began capturing the imagination of executives, business academics, and MBAs. For example, the best-selling books *The Wisdom of Teams* and *Built to Last* were published in 1993 and 1994, respectively.

Professors Matthias Weiss and Martin Hoegl's research supports the argument that America adopted teams relatively recently. They tracked academic research on teams over more than a century, 1902 to 2008. The number of articles written annually remained under twenty till 1974. After rising to about 45 articles in 1980, the numbers took off exponentially—about 135 in 1990, 525 in 1997, and 1,190 in 2006. Largely complementing the discussion above, they wrote:

> In the early 1980s, the teamwork concept experienced a real comeback, entailing a period of growth. A main driver of this development can be found in management innovations. ... Specifically, teamwork was implemented in areas traditionally characterized by individualized and more hierarchical work processes, such as in ... production plants. ... The initial success ... spread and many organizations aimed to adopt it. ... these initial forays ... have led to more team elements ... such as quality circles.[12]

So, by the late 1980s, technologies that empowered people overtook the technologies that had given the Authoritarian Leader dictatorial powers. American and European organizations grudgingly accepted that they couldn't get the best out of teams without fundamentally changing their notions of leadership. The new Empowering Leaders pioneered the ideas we accept as common today.

The leader had to be a coach, a nurturer of people, a solver—rather than a source—of interpersonal and interdepartmental conflict. He—it was still primarily a world of men—had to trust his people and empower them to speak up, even if doing so would result in explicit or implicit criticisms of his decisions.

The explosive growth of business degrees in America and worldwide since 1990 spread these ideas and made teams the building blocks of modern organizations globally. Not surprisingly, 99% of the respondents to the Global Survey (described below) reported belonging to at least one team; 32% reported being on more than five.

The Global Survey also shows the impact of digital technologies. Of these teams, 94% had noncolocated members; 89% had members from other business units of the company; 87% had members based in other countries; and 65% had members who worked for other companies. Leadership in the digital epoch must be compatible with these new realities.

The Emergence of the Digital Epoch

Let's summarize the prior sections. Periodically, technologies appear that have long arcs of impact into the future. When introduced, they require dramatic changes in the nature of work, which, in turn, require profound changes in how people are organized. That changes how people must be led. Companies—and executives—who fail to adapt are cast aside by those who do.

Scientific management and the quality movement gave us long-arc-of-impact technologies—time-and-motion studies, Gantt charts, statistical process control, and other problem-structuring and problem-solving techniques. Scientific management optimized human effort, enabling unskilled people to become employees; quality movement required the analysis of time-phased statistical data, raising education requirements for entry-level jobs.

Scientific management created/strengthened functions and eliminated people's discretion; quality movement returned discretion to empowered people, not individually, but as members of teams. Scientific management created silos; quality movement sought to eliminate them. Scientific management enabled and required the-boss-knows-best, tolerate-no-dissent leadership; quality movement enabled and required the exact opposite—leadership that empowered people.

We are currently absorbing another set of long-arc-of-impact technologies, those of the digital epoch. How are the nature of work, organizations, and leadership changing?

Leadership is rarely a single act. Instead, it consists of mindsets, behaviors, and actions—large and small—in multiple related areas that show up on a daily or weekly or monthly basis. Facing an epochal change, like your predecessors, you too have to rethink leadership.

The Rest of the Book

This book draws on a survey I conducted of 700 midtier to very senior executives, typically from large, multinational organizations. They either attended programs I taught at IMD (about 200) or belonged to a Qualtrics survey panel (500). The distribution of the respondents *roughly* tracked the regional distribution of the headquarters of the two thousand largest global companies: 27% were from the Americas (including Central and South America); 29%, Europe; 6%, Middle East and Africa; 7%, India; 10%, China; 11%, Japan and Korea; 8%, South East Asia, Australia, and New Zealand; and 1%, the rest of world. I'll call this research the *Global Survey* to differentiate it from other cited surveys.

This book also draws on interviews with top executives at select US, European, and Asian companies. I told interviewees that I wasn't interested in how they achieved their positions since their journeys probably began before the digital epoch. Instead, I asked them about attributes they currently considered when deciding that a midtier person had high potential. Those attributes implicitly accounted for the realities of the digital epoch. I interviewed many more executives than I've quoted here; I've identified most, but not all, of the latter. I also interviewed many "high potential" midtier executives; only a few explicitly made it to these pages; none are named. Countless informal discussions with members of both groups generated issues for the formal interviews.

Finally, this book draws on news media, corporate websites, and most of all, the wide body of academic literature on creativity, collaboration, digital (and nondigital) technologies, innovation, medicine, negotiation, and psychology. Specific citations are listed in the Notes.

Part I, "What Do Digital Technologies Really Do?," addresses the seven ways the context for leadership is changing.

Chapter 2, "Restructuring Work and Organizations—Six Principles," discusses why the common description of digital technologies keeps us from understanding their organizational and leadership demands. Instead, it describes six "Principles" that distinguish digital technologies from all prior ones. Akin to those for the scientific management and quality movement epochs, they describe how work and organizational structures have changed.

Chapter 3, "A Digital, VUCA World—The Seventh Principle," addresses one issue not relevant to prior long-arc-of-impact technologies: The impact of digital technologies extends beyond the boundaries of organizations. Leaders must also understand this interaction. Collectively, the seven Principles across the two chapters anchor the subsequent chapters on leadership attributes.

Chapter 4, "Implications for Leadership," discusses six ways in which the Principles are reshaping the context for leadership in the digital epoch. It presents supporting evidence from the Global Survey on the status of these changes worldwide.

Part II, "Leading for Creativity," delves into the mindsets, behaviors, and actions needed to do so.

Chapter 5 is entitled "The Many Faces of Talent." Digital technologies distribute work globally. They also make work cerebral, making physical strength—the reason why men dominated the workforce in centuries past—far less relevant. These realities make inclusionary leadership an existential need for companies.

Chapter 6, "A Broad Wingspan, Not a Long Tail," asserts that today's leaders have to navigate the "in-between spaces" that experts avoid. Executives who know much about narrow areas follow the twentieth-century norm. Digital technologies require leaders with breadth of knowledge, experiences, and skills.

Chapter 7, "Being Truly Collaborative," argues that win-win has become a buzzword used to hide outcomes that benefit one party but don't seriously hurt the other. A world in which work is distributed requires truly collaborative networks of people and organizations. Without them, leaders get cut off from essential knowledge and expertise.

Chapter 8 focuses on "Championing Creativity." In the predigital epochs, neither personal creativity nor the ability to inspire creativity was essential to lead. Today, work is more cerebral and demands attention to emergent needs. So creativity is moving to center stage. Leaders need to embrace empathy, shun the pursuit of uniformity, and acknowledge that an average person with a computer can outperform a smart person without one.

Part III, "Guide Rails for Creative Efforts," addresses how to ensure bold digital technology powered initiatives do not self-destruct.

Chapter 9, "Defining the Uncrossable Lines," addresses a twenty-first-century reality: a never-ending stream of technology-driven, *self-inflicted* crises companies are experiencing around the world. It argues values matter in the digital epoch. Instead of endorsing any specific one, it discusses the process by which leaders should develop and deploy values.

Chapter 10, "Defining Your Strategic Intent," notes that while digital technologies are evolving rapidly, CEOs are making decisions incrementally. To act boldly and effectively, leaders must understand the Five Assumptions they implicitly make about digital technologies: benevolence, infallibility, controllability, omniscience, and authenticity. They must also protect against avoidable errors that can derail major initiatives.

Part IV, "Where Next?," deals with adopting lessons from this book.

Chapter 11, "Building a Personal Leadership Philosophy," describes a process for creating a "personal leadership philosophy" and prioritizing the changes you need at this point in your professional life.

1 What Do Digital Technologies Really Do?

There are more things in heaven and earth, Horatio, than are dreamt of in your philosophy.

—William Shakespeare, *Hamlet*, Act 1 Scene 5

2 Restructuring Work and Organizations—Six Principles

The English system required precise measurements of what was being produced, the American system required rough compatibility of mating pieces, scientific management required the measurement of people's work, and the quality movement required measurement of work over time. These seemingly inconsequential changes modified organizational structures, overturned the prevailing leadership paradigms, and fomented societal turmoil. Following this logic, we need to understand the six key ways digital technologies are changing work and organizational structure.

Words Matter

Suppose a crossword puzzle clue, "Digital," required a ten-letter response. With no intersecting words to guide you, what would you write? Components? Consortium? Interfaces? Leadership? Satellites? Smartphone?

You'd instantly write "DISRUPTION."

Regardless of who you are, where you live, or what you do, for over two decades you have been bombarded by stories about digital disruption. Your word choice would be an inevitable "Don't think of a white bear" moment. As a pathbreaking psychological study established, once told that, it's hard to stop visualizing white bears.[1] Furthermore, words implanted in our brains shape how we think of complex issues.[2]

Words matter.

We have long known that radically new technologies originated in niche markets, grew, and destroyed entire industries.[3, 4] Steamboats, for example, initially couldn't cross seas. So builders of ocean-going sailing ships didn't see them as threats until it was too late.

Clayton Christensen refined this view. If a new technology, however novel, immediately offers benefits to the existing customers of an industry, it gets adopted. He called such technologies "sustaining." In contrast, "disruptive" technologies initially offer benefits that existing customers don't value. So industries ignore outsiders who offer these to niche markets at lower costs. The industry's customers switch *if and when* one or more characteristics improve and become valuable to them. By then, it's too late for the industry to respond. For example, early transistor radios were small, light, and portable but had poor sound quality.[5] As sound quality improved, the other attributes and lower cost made them appealing to everyone.

Almost exactly twenty years after the original article, Christensen argued that overuse had distorted the meaning of disruption beyond recognition:

> [T]he theory's core concepts have been widely misunderstood and its basic tenets frequently misapplied.…too many people who speak of "disruption"…use the term loosely to…describe *any* situation in which an industry is shaken up and previously successful incumbents stumble.…If we get sloppy…the theory's usefulness will be undermined.[6]

Words matter. Seemingly, every digital innovation is now—wrongly—called disruption. As an example, Christensen criticized its application to Uber. Uber neither pursued a lower cost opportunity that existing taxi companies had overlooked nor targeted people who weren't being served. Moreover, it didn't need to improve any feature to attract traditional taxi users. So Uber was a *sustaining* innovation.

Of the countless other mischaracterizations, consider two common ones:

- During the period 2010–2017, venture capitalists invested $170 billion—more than the combined 2017 gross domestic product of all but 18 of the world's 211 countries!—in "fintech" (financial technology) start-ups.[7] These haven't threatened any major financial institution globally; every change in the rankings of the top twenty-five global banks has been due to China's rise and the Great Recession.

Banks have rapidly absorbed, reshaped, and even created new fintech tools.

• While Airbnb is a down-market offering that expanded the market, it can't improve any feature that hotel users desire. Hotels' occupancy rates have remained stable and high, while their average daily room rates and the number of days when they were full have risen every year since 2009.[8] Large cities have filed lawsuits against, and imposed hotel-like taxes on, Airbnb and its listers, reducing its cost advantages.[9] Objectively, Airbnb is an example of a "blue ocean strategy," not disruption.[10]

Moreover, Airbnb and Uber are merely unregulated traders of real assets that others own. They are no different from commodity traders listed on the Chicago Board of Trade since 1848. Commodities traders have always been richer than farmers, but without farmers, they wouldn't survive a day.

Words matter. They shape our thoughts, just as much as our thoughts shape them.

Christensen was rightfully concerned that the misuse of disruption is reducing its strategic value. Far more importantly, it is distorting the focus of leadership in the digital epoch.

Disruption has become a buzzword that keeps leaders from understanding how digital technologies—like prior long-arc-of-impact technologies—actually transform work and organizations. Asking the wrong questions, they neglect the bold investments, the new partnerships, and the creativity they need to enter new markets, cure cancer, or build a sustainable economy.

Equally importantly, calling every innovation a disruption turns each into a jingoistic existential crisis that only hard-edged warriors can fight. Whom can we destroy? Who could attack us? Which customers can we keep 100% loyal? Such objectives also create destructive work environments by excluding those who don't fit the stereotype of combatants. The inevitable result is the misogynistic "bro" culture that Silicon Valley is belatedly and half-heartedly tackling.[11]

Words matter. It's time to sharply limit the term "digital disruption" to instances where it's actually happening.

An Extended Example of Digital Transformation:
The Pharmaceuticals Industry[12]

Digital technologies, like all long-arc-of-impact technologies, change the basic questions—what, why, who, how, when, where, and how much—which govern the work people do. They positively shape *all* organizations—small or large, existing or start-up—that respond appropriately and destroy those that don't. The pharmaceutical industry offers key insights into this process.

Prior to the digital epoch, individual pharmaceutical companies used to develop drugs, often building on the work of, or in collaboration with, academic scientists. Large "discovery-and-development" teams of their scientists focused on specific disease areas. They knew the disease, but not necessarily what caused it (the "target").

Because of this uncertainty, years of often-frustrating experiment design and trial-and-error testing in animals followed. Two questions guided this effort:[13] What happens? How can we produce the intended impact? The pharma scientists worked to identify the types and shapes of the many possible targets. They also extracted or created molecules of the active ingredients that could be used to formulate drugs. The tools they used came from specialist companies, but the experiments they performed were proprietary and manual. Between 95% and 97% of the drug options they tested failed.

When a drug was found, very large human trials followed. The testing checked not just for efficacy ("Does it cure?") but toxicity ("Does it do any unanticipated damage?"). Even so, some types of toxicity could only be identified when the drug was actually being used. The drug worked for the average patient but didn't necessarily benefit everyone. The overall effort took twelve to fifteen years and cost $1–$1.5 billion.

The first companies to create genomic drugs appeared in the 1970s. They focused on a handful of genes (the "targets") known to cause specific illnesses, such as the inability to secrete insulin. These isolated, pioneering efforts foreshadowed the digital epoch's efforts.

In 1990, the United States began funding the Human Genome Project, an ambitious effort to map human DNA.[14] The explosive growth in digital technologies during the 1990s led to the creation of very precise tools for sequencing, cutting, and splicing genes. Consequently, the project began delivering key results in about ten years, five sooner than originally expected. The draft human genome was produced in early 2001, the near-complete genome was produced in 2003, and the remaining bits of work were completed in 2006. The age of digitally powered genomic medicine had arrived.

Now, in several therapeutic areas of drug development, scientists know from the start why the disease occurs. They focus on specific subsets of patients afflicted by the targeted gene.[15] Pathways the disease takes are established early in the process, allowing subsequent experimentation to take less time. Only 60%–65% of the drug candidates fail. Human clinical trials can be shorter. Toxicity is predictable and addressed during drug approval. The overall effort costs $0.5–$0.8 billion and takes less than ten years.

This profound change also restructured the industry: Work got distributed across companies around the world.[16] Academic researchers now play comparatively smaller roles. Relatively recent start-ups—genetics-focused biopharma companies and contract research organizations—play comparatively larger ones. Pharma company discovery-and-development teams are smaller and more focused. They design specific tests, and biopharma scientists execute them using tissue samples with the desired genetic characteristics that are sourced by contract research organizations from around the world. The scientists collaborate with toolmakers to develop the tools they need. Their experimental work is largely automated and uses standardized kits.

Despite the many start-ups, the list of major pharma companies has remained largely unchanged since 2003. However, all established firms do most tasks differently and don't do at all some of the tasks they used to do. Industry entrants do new work that was either a small part of old pharma's work or not needed before genomic medicine. Working with existing firms, whom they don't threaten, they are now essential for success.

Other digital efforts accompanied these changes. Johnson & Johnson worked to connect digital knowledge bases so different groups of scientists located across its many operating companies could better share their research findings.[17] Recognizing that it couldn't be overly dependent on blockbuster drugs while the entire health-care system was changing, Boehringer Ingelheim deployed digital tools that networked a broader set of employees and facilitated stranger-to-stranger contact.[18] It even established a virtual school of intrapreneurship. Novartis transformed its sales force's interactions with doctors by replacing all paper- and laptop-based material with iPad-based interactive tools. This required extensive rewiring of internal knowledge bases as well as changes to performance evaluation systems and even business models.[19]

These examples are about changes within individual companies. Like the shift in drug discovery, bigger changes will require working across companies.

In 2002, concerned with the rigidity of the US drug distribution system, GlaxoSmithKline (GSK) invited senior executives of wholesalers (McKesson, Cardinal Health, AmerisourceBergen) and retailers (like the giants CVS and Walmart) to a symposium. (Disclosure: I organized this event.)[20] In the digital epoch, medicine would inevitably require sending medicines to, and collecting data from, individual patients. If radio-frequency identification (RFID) chips could be safely embedded to track individual pills from manufacture to actual consumption, how could the practice of medicine change?

Many patients don't take all their medicine. Among other things, this enables serious diseases to become drug resistant. RFID could ameliorate this problem. More immediately, it could help the industry with counterfeiting, tracking and tracing for possible recalls, and multiple other challenges.

The industry first had to abandon a long-standing business model. Pharma companies used to predictably raise their prices, and wholesalers and retailers used to buy drugs at the "old" prices and sell at the "new" prices. This inflated their (nonoperating) "trading profits" while protecting inefficient wholesalers and retailers from market discipline. Already threatened by the Sarbanes-Oxley Act, this model would impede the flow of essential data. After an animated discussion, the executives concluded that the implementation of a digitally supported distribution capability would require more than ten years.

Months later, the industry began dismantling its archaic business model. A collaborative effort then enabled the adoption of electronic product codes on packaging but stopped well short of what was truly needed. There were some key technical hurdles, but business issues, including sharing of costs, value, and data, had been the principal impediments. In 2019, regulatory deadlines for drug security are looming and the industry still has much to do.[21, 22]

The US Food and Drug Administration (FDA) had approved the first RFID tag for human implantation in 2004.[23] The American Medical Association had issued its code of ethics for RFID implants in 2007.[24] In October 2017, the FDA approved the first drug with an embedded chip, for schizophrenia.[25] When consumed, it signals a smartphone, which forwards the data to designated health-care providers. Going beyond one limited-use drug to real-time interactions with many patients consuming many such drugs will require a collaboration-and-experiential-learning-driven transformation effort comparable to the adoption of gene-based drug discovery. Seventeen years after the GSK conference, that challenge is still not an industry priority.

What Digital Technologies Actually Do

In science and engineering, "first principles" are axioms, equations, or ideas that are known truths. Building on them produces more reliable answers than building on expedient facts or data. Applying a similar logic to the pharmaceutical industry—and looking for supporting evidence and counterevidence in other industries—led to the identification of six Principles that govern how digital technologies are restructuring work and organizational structures. These, together with a seventh Principle discussed in chapter 3, differentiate digital technologies from all prior ones.

The Principles are relatively self-explanatory; the discussion below largely relies on examples to establish their relevance. That said, being subject to human actions, they—unlike the laws of physics or mathematics—are not inviolable. It is possible cite examples where a Principle should, but does not, hold. Even so, they are changing the context within which you must lead.

Principle 1: Digital technologies reduce, or eliminate, the value of an elite group's skills or knowledge and enable—and may even require—the automation of its work (see also chapter 10).

The 1980 Nobel Prize in Chemistry recognized the development of an efficient, manual way of setting up genetic tests. This key skill of genetics pioneers improved incrementally over time, until digital technologies automated it completely. Digital technologies also enabled the simultaneous—not sequential—study of chemical molecules and biological structures, sharply reducing costs. Consequently, today many digital native consumer businesses use cheap test kits to test people's ancestry or make personalized recommendations for wine.

The following are examples from other industries:

- For almost two hundred years, London hackney/taxi drivers have been renowned for their ability to take passengers to obscure addresses by the shortest or fastest paths. Passing the grueling "Knowledge" licensing exam required excellent memory and spatial thinking skills.[26] Starting in 2005, GPS systems began degrading the value of these abilities.

- In 2012, the Hewlett Foundation set up a challenge for automatic assessment of essay answers.[27] Driven by artificial intelligence (AI), these systems are getting better, and their use is rising.[28] They can sharply reduce the time needed to manually grade assignments, freeing teachers to focus on creative activities like coaching the essay-writing process.

- In 2016, IBM's AI engine Watson diagnosed some cancers more accurately and quickly than human doctors. At a University of Tokyo teaching hospital, it diagnosed a rare leukemia.[29] At a research hospital in Bengaluru, India, it recommended a treatment that obviated a double mastectomy that surgeons would otherwise have recommended.[30]

Principle 2: Digital technologies augment the capabilities of less skilled people, enabling them to undertake tasks they couldn't earlier (see also chapter 10).

As some people lose skills, others gain complementary ones. Many pharma companies used to divide gene-based drug discovery into "dry science" (designing genetics experiments and analyzing digital data) and "wet science" (traditional lab-based research with tissues and animals). Individuals with PhDs did the former, and those with MS degrees the latter. Advanced digital technologies are blurring this line, broadening the responsibilities of the latter group.

Examples from other industries include the following:

- By 2009, GPS-equipped smartphones gave average Londoners a capability their licensed taxi drivers didn't have: forewarning of congestion, accidents, and road closures. A good driver with a smartphone could navigate better than a great taxi driver without one.
- Advances in neural machine translation technology[31] are enabling far better any-language-to-any-language (near) real-time translation than was available just a few years ago.[32] A health-care version recently got an encouraging evaluation for translating emergency room discharge instructions from English to Spanish and Chinese.[33, 34]
- Design engineers now have computer-aided design (CAD) tools that they can use to iteratively generate many ways of achieving predefined design goals.[35]

Incidentally, "less skilled" doesn't necessarily mean "less educated." Personal computers gave executives the ability to produce documents they couldn't have done formerly. Secretaries skilled in shorthand and typing are no longer needed. Downloadable images and templates built into widely used presentation software have almost eliminated the need for professional graphic artists.

Principles 1 and 2 can act independently and will do so increasingly often. Drones that survey transmission lines or pipelines in remote areas sharply upgrade the job content of surveyors. Instead of spending days, even weeks, in the field, computer-equipped surveyors in offices receive, and semiautomatically act on, diagnostic reports the drones send. This upskilling is reducing the numbers of surveyors needed. Similarly, driverless trucks will eliminate many jobs within a few years.[36] New jobs will emerge but current drivers may not benefit from these.[37]

Principle 3: Digital technologies enable—and may even require— work to be distributed over time and geography (across outposts of a company and across companies).

Drug discovery work is now spread across multiple companies around the world. A pharma company scientist designs a genetic test for cancer. A network of oncologists affiliated with a contract research organization halfway across the world provides the tissue samples with the necessary genetic diversity. The biotech company scientist sequences and analyzes genes in the tissue samples. The pharma scientist then receives the results of the executed tests, and the symbiotic cycle continues.

Global companies in other industries are internally networked as in the Johnson & Johnson and Boehringer Ingelheim examples. Consulting companies staff project teams with people based in different locations. Technology service companies "follow the sun," handing over work from one office to another in order to assure round-the-clock system reliability for their clients.

Moreover, established companies compete not as independent entities but as interdependent members of networks. Initially created to outsource commodity work—manufacturing or back-office support— and reduce costs, these networks took on intellectual property creation when CAD and digital coordination of work improved. Examples of the results of these changes include the following:

• Five European, four Asian, and four North American companies cocreated the Boeing 787's hardware.[38, 39] As in the past, Boeing built 30% of the aircraft and suppliers 70%. However, their basic approach to design changed. The companies collaborated for over eighteen months to select advanced composite materials they hadn't used before. Each then engineered and built their assigned sections of the aircraft in different parts of the world and shipped these to Boeing. Each also had a financial stake, having funded the research and development (R & D) it needed for the aircraft. Incidentally, Boeing didn't pioneer this approach; its chief rival, Airbus, originally an alliance of four European aircraft companies, did.

- Automakers around the world also work similarly. Substantial portions of cars are designed by companies whose names are not on the car's body. Well more than half the manufacturing is done by them.

- The outsourcing of information technology (IT) services, customer service, reading of X-ray/MRI reports, investment analyses, and countless other activities we experience at work or as consumers are examples of this Principle. Remove computers, and most would simply not have been possible.

- Major airlines "code share" flights with their network partners (e.g., Star Alliance). While anyone can buy a code-share seat, the arrangement allows airlines to hand over their loyal customers to trusted partners.

Principle 4: Digital technologies enable—and even require—work to be increasingly thought driven instead of being muscle powered.

When pharma companies adopted high-speed screening and sequencing, they raised the importance of test design and data interpretation and eliminated the value of manual test setup. Biopharma companies used to set up and conduct tests to screen and sequence genes and molecules. They now regularly develop unique analysis capabilities to attract capital and top scientists.

Elsewhere, digital technologies initially enabled better, faster, or cheaper physical work: By manipulating symbols on computers, people directed machines to cut metal or move an aircraft's ailerons. As computer intermediation became ubiquitous and deeper, it became possible to buy, license, or sell ideas, concepts, and algorithms independently of the outputs they created.

The nature of work changed: Much meaningful work now occurs inside people's head, not outside their bodies. The work—visualizing the future, creating and testing models, writing the code—gets frontloaded. The actual implementation is often effected or aided by computers, though people (still) respond to real-time data and issue needed instructions via computers.

Examples in other industries include the following:

- 3-D printing is already being used to make many complex parts for cars and aircraft.[40, 41] The human work is largely in developing the code; computers handle the execution. Simple houses designed by architects on CAD terminals are being built by very large 3-D printers that can lay down layers of concrete.[42] Current buildings are simple and not yet deployable at scale[43] but are improving; Dubai expects to have 25% of its buildings 3-D printed by 2025.[44] Use of 3-D printing to replace defective organs in humans is in various stages of development.[45]

- The entire field of robotics is an example of this Principle: Robots translate human intent/commands into actual action. Factory robots are taking over much—but not all—work done in manufacturing operations. Over the last decade, robotic surgery is becoming increasingly used in urology, gynecology, gastroenterology, and cardiology. Both factory and surgical robots are often more precise than humans; the former also eliminate the need for physical strength.

- Digital translation of human intent into action is also transforming services. "Software robots" (called "bots"), both beneficial and malicious, are already ubiquitous online and will continue to take over even more human activities. In 2018, Google Duplex, a prototype AI bot, perfectly mimicked human conversation, down to the "ums" and "let me sees,"[46] as it made calls to set up appointments and reservations. "Brain-computer interface" technologies enable a thought and a twitch of a muscle to produce desired action. Already used for some advanced health care, they may find use outside that field.[47]

Because of Principles 3 and 4, international trade in intellectual property—royalties, license fees, and so forth—has skyrocketed. In 1993, twelve countries had signed three international preferential trade agreements; in 2014 the comparable numbers were eighty and forty-five, respectively.[48] Since 2010, exports of intellectual property–related services have grown an average of 7% a year; in 2017, they grew 10%, more than any other commercial service.[49]

Principle 5: Digital technologies create needs that aren't predictable and/or add disproportionately great value.

Historically, while the form of human needs changed, their essence endured. CDs replaced cassettes, but music has remained constant. TVs gave speeded-up access to news and made it visual but didn't inherently change the determinants of quality journalism. Cars replaced horse carriages, but people still used roads. In the service of these "enduring needs"—music, news, travel—new technologies created *predictable* ancillary changes: Cars needed paved roads and gas stations instead of horse stables. In contrast, digital technologies create new needs ("emergent needs") that *aren't predictable and/or add disproportionately great value.* Not pursuing these opportunities can have costly, even deleterious, impacts.

The pharma industry used the long, costly cycles of drug development to rationalize the limiting of testing to Caucasian men, and assuming unproven efficacy for everyone else (including Caucasian women and children). For gene-based medicines, this strategy is suboptimal since scientists must understand what makes particular groups of people different from others. Tissues are now sourced from regions of the world where populations have great genetic diversity.

Twentieth-century pharma scientists didn't need expertise in managing huge databases. Today each genomic experiment generates a terabyte of data. Research teams must also continually refine how they process these data. Regularly developing new ways of manipulating, analyzing, and storing data is a key capability for biopharma firms; without it, novel discoveries would be hard to come by.

As genomic medicines become common, the capability to handle orders of magnitude more (and different) data than the industry can today will become essential. The data standards and infrastructure needed to track and trace drugs won't automatically handle communications between doctors and patients using RFID-embedded medicines.

Emergent needs change the core of the work that must be done. They require new knowledge, new systems, new training, new people and organizations to connect to each other. They can subtly shift power balances inside organizations: Corporate information technology departments used to support most technology needs for drug R & D teams; today, data scientists with PhDs are full members of the R & D teams.

Examples in other industries include the following:

- Operating systems don't provide all of a digital device's capabilities; apps strike off in unforeseen directions. BlackBerry and Nokia failed in large part because they underestimated the power of the Android and iOS app stores. These allowed individuals and companies to develop solutions to a myriad of niche problems at a scale no smartphone manufacturer could match. Wedded to their existing knowledge bases and power structures, neither company built the knowledge and alliances they needed to serve the new emergent needs.

- The emerging Internet of Things is posing a similar challenge. Many companies are creating new offerings. No matter how brilliant, to take off, these must connect to the "Internet of Things backbone" (comparable to a smartphone's operating system). Conversely, many companies are racing to build the Internet of Things backbone to support diverse, custom plug-and-play tools (comparable to apps) that will serve countless unpredictable needs.[50]

Principle 6: Digital technologies expose organizations to radical transparency, which may—or may not—benefit them individually or their networks or society at large.

The US pharma industry's inexcusably flawed current business model is nevertheless an improvement on the egregiously opaque one it abandoned roughly a decade ago. That model assured profitability for even the most inefficient wholesalers, distributors, and retailers. The electronic product codes–based model has a long way to go, but it is enabling the industry to identify drug diversions or contamination, assure drug freshness, and initiate any needed recall. Society benefited, though weak wholesalers, distributors, and retailers, forced to give up a profitable business model, undoubtedly suffered.

It is hard to find examples of industries that *aren't* affected by this Principle. Searching online has become ubiquitous and essential for modern life. Privacy issues have triggered both mass surveillance of people[51, 52] and Europe's "right to be forgotten" laws.[53] Consumer data breaches have become so large and frequent that they have to be truly horrendous to draw attention. Actions of organizations and governments that are opposed by others, whether legitimately or not, are hard to keep secret.

A single research study puts in perspective the importance of radical transparency. The automated analysis of three hundred or less arbitrary "likes" on Facebook accurately predicted "a wide variety of people's personal attributes, ranging from sexual orientation to intelligence."[54] The power of this Principle shouldn't be taken lightly.

De-skilling (Principle 1), upskilling (Principle 2), cerebral work (Principle 4), emergent needs (Principle 5), and radical transparency (Principle 6) are changing the nature of work. Distributed work (Principle 3) is transforming organizations by making internal and external networks essential for success.

Two Principles in particular—cerebral work (Principle 4) and emergent needs (Principle 5)—make creativity critically important in the digital epoch; the other Principles play supporting roles. Businesses that ignore these and focus intensely on Principle 1 will remain anchored in the logic of past epochs until it is too late.

Leadership is about making choices and pursuing difficult ends. Executives who stick to well-hewed paths may manage well, but they forfeit the right to be called leaders.

3   A Digital, VUCA World—The Seventh Principle

The six Principles in chapter 2 capture how digital technologies, like prior long-arc-of-impact technologies, transform work and organizational structures. Digital technologies also do something that no prior long-arc-of-impact technology ever did: **Principle 7: Digital technologies interact with and affect an organization's external environment**. This Principle is also reshaping the context for leadership.

A Volatile, Uncertain, Complex, and Ambiguous World

On November 9, 1989, the Berlin Wall began coming down. It had symbolized the Cold War, the great divide between the US-led North Atlantic Treaty Organization nations and the Soviet Union–led Warsaw Pact that shaped the second half of the twentieth century.

Humanity had weathered an existential risk. Though conventional wars between "client states" killed thousands worldwide, guided by the game theory–based concept of mutually assured destruction, the two alliances avoided global annihilation. The world started moving toward peace, with occasional upheavals like the genocide in Rwanda and the bloody dissolution of Yugoslavia.

But reality is never that simple.

The bipolar world became multipolar. The European Union could reject US interests when it disagreed. Russia still wielded great nuclear power and vast energy resources. East Europeans clamored to join the European Union but resisted its ideals. China grew rapidly economically and militarily but rejected global trade norms. Japan hit a wall economically, but its companies dominated key industries. South Korea became an economic powerhouse. Sleepy Association of Southeast Asian Nations (ASEAN) countries became "Asian tigers." Nuclear armed India finally started liberalizing its economy. South Africa shed its apartheid past and rejoined the global economy. Argentina, Brazil, and Chile emerged from right-wing military dictatorships and became messy democracies with erratic economies.

The struggle for resources grew acute and complicated. Changes humans themselves had unleashed threatened humanity's common future: Climate change began accelerating dangerously. Humanity could no longer ignore balancing needs and desires—digital devices required rare earths that are mined in regions mired in people-created disasters. Independence movements roiled formerly stable countries, even within the European Union. Terrorism, for long "their problem, not ours," afflicted countries globally and triggered never-ending wars.

Goals became indistinct, and reality even more so. International relations ranged from durable to fleeting, but the latter regularly won. Important work couldn't be done alone, but gaining other nations' support could be very hard.

A stream of individually small but collectively difficult problems began fomenting at least as much fear as had the prior well-understood existential crisis. In 1987, the US Army War College, anticipating these geo-socio-political changes, coined an acronym, VUCA (volatile, uncertain, complex, and ambiguous), to capture their essence.[1]

A VUCA Business World

Businesses co-opted the acronym around 2003 as the economic implications became obvious. New businesses were making industry boundaries so murky that multiple governments changed long-standing systems for tracking economic data. Digital technologies were creating undreamed-of opportunities and enabling the global ambitions of tiny firms but were also posing unprecedented challenges. New competitors were rising far from the traditional power bases.

With the passage of time, VUCA has become another buzzword: Like "disruption," it gets tossed around so much that it has lost its meaning.

It gets added to strategy presentations but demands no change in mind-set, behaviors, or actions.

Type "VUCA and leadership" in a browser, and many sites, including several fronted by business professors or retired generals, will pop up. Most suggest familiar business concepts that begin with the same letters as antidotes for VUCA. Such advice is facile, even specious: A brilliant five-year vision, a supposed antidote for volatility, won't help resolve today's completely unrelated volatile situation. Conversely, that vision may perish if the volatility isn't successfully addressed.

Digital technologies can create and exacerbate VUCA conditions. Appreciating why requires understanding its constituent terms. The context of a single industry—let's consider energy—reveals how sharply they differ.

V, for volatility, refers to unpredictable changes that unfold very quickly. When fear and greed run rampant in the immediate aftermath of ter-rorist attacks, oil prices fluctuate wildly. Volatility can potently affect the course of events because executives often can't specifically antic-ipate what event will trigger volatility when, where, how, and why. So they must decide in real time how, indeed if at all, to respond.

U, for uncertainty, refers to foreseeable variation within known limits. Global energy use typically varies predictably with the level of eco-nomic growth. Endemic in nature, uncertainly can't be eliminated but can usually be modeled. Business leaders, however, rarely use these models: Most people, even those trained in statistics, lack an intuitive understanding of its implications.[2] Nor do their organiza-tions: Single point measures shorn of uncertainty (e.g., specific profit targets) are easier to deploy, even if they wrongly imply that the future will resemble the present.

C, for complexity, refers to the many interacting factors that shape an issue. Complex systems make shocks hard to manage. Stopped at any point, shocks propagate by unanticipated paths. The electricity grid for the Eastern United States is one of the world's most complex. In 2003, fifty million people spent two days in darkness while experts struggled to understand why it had suddenly crashed. Psychological research shows that a human limitation aggravates this problem: Our brain, unable to make sense of many factors (particularly if imposed all at once), grossly simplifies cause-and-effect rationalizations and reaches suboptimal conclusions.[3]

A, for ambiguity, refers to the possibility of no good answer, or multiple possible answers without strong support. Many knowledgeable, rational people believe nuclear energy can help humans respond to climate change. Equally knowledgeable, rational people disagree, noting that fissionable materials like uranium and plutonium are hard to store safely. Both groups can cite tons of supporting data but won't convince each other because their positions are driven by their deeply held values.

A Digital, VUCA Business World: An Extended Example

The US financial services industry, the epicenter of the 2008–2009 recession, often called the Great Recession, that affected many countries worldwide, is a perfect example of the interactions of digital technologies and the VUCA world. Unlike conventional analyses based on politics or economics, this argument draws on research from computer science,[4, 5] risk management,[6] operations research,[7] and finance.[8, 9] Two stories readily convey their lessons, which essentially address how shocks flow from point A to point B in computer networks.

In 1983, I, a newly minted MBA, joined a global bank at its headquarters near Wall Street. With no prior work experience, I earned perhaps the lowest salary in my graduating class and wasn't eligible for bonuses or stock options. Yet, I was sure I knew more than everyone else.

My bosses tolerated my arrogance. Three months into the bank's year-long management training program, they gave me tasks that my peers didn't get for years. Supervised by a senior officer (who did get bonuses and stock options and was in line for a career-making promotion), I developed position papers for our board of directors. Meant to be starting points for discussions on new policies for risk tolerance and lending standards, they were usually approved without major change.

My bank entered the nascent market for very high-risk, high-reward loans to commodities brokers. My bosses knew little about this market, and I, nothing. The newly hired, hard-charging lenders who specialized in that industry pooh-poohed obvious risks. Even so, my bosses supervised me with a lighter touch than usual. The lenders' inputs overly influenced my analyses, and my bank quickly approved large, dubious loans. Much money was at stake for many people.

Did many of those loans become troubled? I can guarantee that. Did my bank lose money on them? Almost certainly! Were other institutions taking similar risks? Since the regulators and markets considered us rock-solid while our leaders were being imprudent, I have no doubts we had good company.

But my failures didn't bring down my bank. My bank's shortcomings didn't radiate across the industry. And collectively, my generation of young know-it-all financiers didn't destroy the world economy.

Why not? We assessed investments individually, using handheld calculators and primitive Lotus 1-2-3 spreadsheets. The ability to bundle and sell off our loans didn't exist. Our banks' finances weren't intertwined electronically with those of other institutions. Had our bad decisions threatened others, the regulators would have easily walled us off. Run on primitive digital technologies, our world wasn't VUCA.

Fast forward to 2008. Ask yourself, "How did America's subprime mortgage crisis cause Iceland's three major banks to simultaneously go bankrupt?" By then the financial services industry was the most digitally interlinked one anywhere. Its unidentified VUCA conditions baffled its own executives and the regulators.

Cerebral work enabled esoteric digital products and services whose risks few understood. Digital distribution of risk via (instantaneous) interinstitutional investments allowed localized problems to spread with lightning speed. (As the crisis unfolded, the grossly overleveraged Icelandic banks, with huge exposures [mostly in other European countries], lost access to credit markets.[10]) Automated decision-making produced great volatility: Share prices for even sound institutions plummeted and contagiously spread fear when mere hints of problems at institutions like Lehman Brothers and Bear Stearns set off computerized trades. Incomplete knowledge, adequate for the predigital epoch but no longer, led many financial institutions to invest on both sides of failing transactions and unknowingly create lose-lose conditions. These destabilized institutions like AIG, which had insured these loans.

Complexity stymied the US Federal Reserve Bank and the US Treasury's initial efforts to contain the crisis by sequentially rescuing weak institutions. They succeeded only by resorting to the much-criticized industry-level bailouts; these shored up multiple weak institutions and buffered stronger ones that bore systemic risk. Doing so, however, created a perfectly ambiguous "moral hazard" problem: Would the bailouts increase risky behavior by creating expectations of future bailouts?

Undoubtedly, greed and deceit abounded, or flawed, esoteric products that were sold only to the less sophisticated wouldn't have been created. But greed and deceit existed in 1983 too, when my employer's unwarranted risks didn't destroy the world's economy. Blaming greed and deceit *alone* is a gross oversimplification.

The Shape of Things to Come

Despite the bigger "shock absorbers" created by the regulatory changes made since then, the fundamental risks haven't diminished. Computers are faster (volatility), metrics used to measure risk tolerance are still static (uncertainty), and networks are more extensive (complexity). The pesky moral hazard problem (ambiguity) won't go away.

Other industries now apply similar digital technologies extensively; they can foment crises that will make 2008 seem like child's play. The burgeoning Internet of Things is connecting ever more products and services to other products and services, systems to other systems, and companies to other companies. These will inevitably challenge our ability to diagnose and fortify failure points. Small, seemingly inconsequential failures will carry within them the potential to harm not just a company but the economy.

This isn't an exaggeration. Heavy snowfall in one US state and *one* bug in key software triggered the aforementioned collapse of the Eastern US electricity grid. The snow brought *first one, then three more* high tension wires in that state (out of the many hundreds there) in contact with trees, and the software didn't detect the problem.[11]

Despite these potentially huge costs, the use of digital technologies will grow. They offer many benefits, and most people in most places would be worse off without them. They helped create both the Boeing 787 and life-saving drugs. They allow companies to assemble project teams with talent from around the world. They provide phone-based banking services to poor people ignored by traditional banks. And they may someday mitigate hunger, cure cancer, and help deal with climate change. In any case, no one in history has ever been able to stop the progress of any major technology. There is no reason to believe digital technologies will be stopped either.

Next, we turn to how these seven Principles are changing the context for leadership in the twenty-first century.

4 Implications for Leadership

Individually and collectively, seven Principles that explain what digital technologies actually do are profoundly changing the context for leadership. The Global Survey showed that respondents worldwide typically believed their own leaders were not responding adequately to these changes. This chapter describes the new context.

The Seven Principles (see also chapter 10)

1. Digital technologies reduce, or eliminate, the value of an elite group's skills or knowledge and enable—and may even require—the automation of its work.

2. Digital technologies augment the capabilities of less-skilled people, enabling them to undertake tasks they couldn't earlier.

3. Digital technologies enable—and may even require—work to be distributed over time and geography.

4. Digital technologies enable—and even require—work to be increasingly thought-driven instead of being muscle-powered.

5. Digital technologies create needs that aren't predictable and/or add disproportionately great value.

6. Digital technologies expose organizations to radical transparency, which may—or may not—benefit them individually, or their networks or society at large.

7. Digital technologies interact with and affect an organization's external environment.

The Pressure for Culture and Gender Equality Is Rising

Leadership education offered by business schools, leadership solutions promoted by consultants, and leadership policies implemented by most companies have long made three key, linked assumptions. The first two are these:

Most people working together are governed by the common organizational culture, policies, processes, and structures of a single company. Since each company is relatively self-contained, executives can assume a well-understood, shared framework for their leadership initiatives. As long as others (e.g., suppliers) adhere to legal contracts, how their people are led doesn't matter.

Most employees share a handful of cultural, lingual, religious, racial, and political heritages. Home-base and similar countries provide companies, even large multinationals, with most of their employees and customers. Therefore, most employees fit into a few categories, and leaders guide people who mostly look like them, talk like them, pray to the same gods, eat the same food, and so on. Business units located elsewhere rely on equally homogeneous local employees. Their leaders graft essential elements of the local culture onto the home-base corporate culture.

These two assumptions largely held true throughout the twentieth century in the world's largest economy. So they influenced American scholars who have long dominated business education. They spread elsewhere as Americans helped set up business schools in Europe and Asia and exported teaching material worldwide. America also awarded more business doctorates than any other country and published most of the top business journals. Thus, its experiences became "received wisdom" everywhere, even when they didn't quite capture the local context.

Digital technologies make these assumptions unrealistic. By distributing work (Principle 3), they make people half a world away critical to success. They may be from very different cultures and work for organizations with different policies. Radical transparency (Principle 6) may instantly taint a brand name globally by bringing to light problematic conditions in a company's partner.

Of the executives who responded to the Global Survey, roughly 40% each (totaling 80%) "Strongly Agreed" or "Agreed" with the proposition

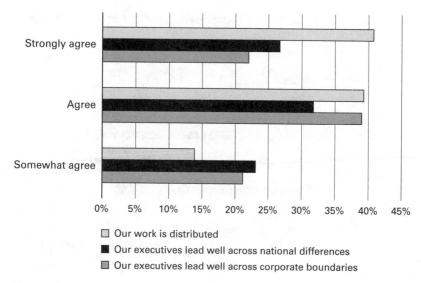

Figure 4.1
Globally, when work is distributed, executives don't lead well across national differences and organizational boundaries.

that their work was distributed across space and time (see figure 4.1). At the same time, they gave relatively tepid support for how their own executives performed in the light of this reality. "Strongly Agree" responses dropped roughly by half (22%) and the "Somewhat Agree" ones rose commensurately in their assessments of their executives' effectiveness at leading across corporate boundaries. Only 27% "Strongly Agreed" and 32% "Agreed" that their executives led effectively across national boundaries. The roughly 20% combined drops in response to both questions reveal a major weakness.

Roughly twenty years after the emergence of the digital epoch, the average ratings in the Global Survey for leading across national differences and corporate boundaries were dismal around the world (see figure 4.2). Data from regions with more than 50 respondents show that none even got close to "Strongly Agree" on these two questions. India's relatively stronger rating for leading across corporate differences is probably due to the outsourced knowledge work done there. Continental Europeans rated their companies most harshly. Their worse than "Somewhat agree" scores could indicate either a regional weakness or,

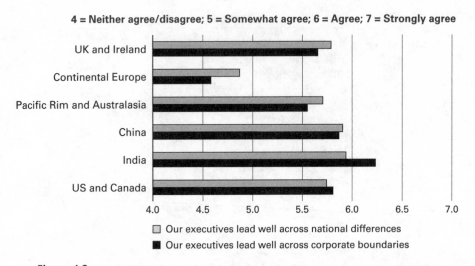

4 = Neither agree/disagree; 5 = Somewhat agree; 6 = Agree; 7 = Strongly agree

□ Our executives lead well across national differences
■ Our executives lead well across corporate boundaries

Figure 4.2
Worldwide, executives aren't leading effectively across national differences and organizational boundaries (for regions with *n* > 50).

what is more likely, a greater degree of self-awareness. If the latter, the responses from other areas are optimistic. In any case, globally, leaders have much to improve.

Research shows that individuals from around the world favor leadership styles that are consistent with their cultures.[1] Even today, CEOs are mostly native-born—or occasionally, naturalized—citizens or from similar and/or neighboring countries.[2] Western leaders in Pacific Asia believe their local executives are less prepared to succeed them than executives from elsewhere. So those Pacific Asian executives are more receptive to calls from executive search firms.[3] They are also more skeptical of their headquarters than are Asian executives in Asian companies. However, this isn't just a Western problem.

Japanese companies rely heavily on expats (Japanese executives sent abroad),[4] having failed to integrate foreigners (foreign executives working overseas in Japanese companies) into leadership positions.[5] Top Japanese executives know that this failure is hurting their businesses. One interviewee, the senior-most HR executive in South East Asia, had prepared an eighty-page analysis for his corporate officers. He lamented that highly prized fluently trilingual (in Japanese, English, and Mandarin) or bilingual (in Japanese and English) foreigners regularly refused job offers, expecting their nationalities to stymie their progress.

The third assumption implicitly made about leadership is this: *A company's leadership standards embody the worldview of its dominant executives.* Most executives in leadership positions come from the home-base (or similar) country. Executives from underrepresented backgrounds, or other nations, adapt and conform, or they forfeit the right to lead. The dominant group is overwhelmingly male.

This assumption, unlike the prior two, isn't solely attributable to America; bias and patriarchy are present all around the world. In most places, women have had to put up with behavior ranging from unconscious slights to full-on abuse.

Creditably, American and European countries collect and discuss data on this issue. These show they have a long way to go. Among US S&P 500 companies, "fewer large companies are run by women than by men named John."[6] In 2018, only 4.9% of CEOs leading European and American companies were women.[7] As a result, many hiring decisions end up enforcing the status quo and perpetuating the misguided notion that the traits associated with being male define "good leaders."

Digital technologies are forcing—and facilitating—a rethinking of this assumption, too. Meaningful work is increasingly thought driven and happens inside people's heads, not muscle powered and happening outside people's bodies (Principle 4). Throughout history, men have held power in workplaces globally because they were, on average, physically stronger. This advantage is increasingly irrelevant. Moreover, radical transparency (Principle 6) can expose discriminatory behavior worldwide, putting at risk the ability to attract talented people. As a result, young people, women, minorities, and foreigners are in positions to challenge these anachronistic ideas or withhold intellectual contributions by simply walking away.

Chapter 5 discusses these issues, gives examples of companies that are redefining leadership to make it inclusive, and suggests what individuals must do.

VUCA Conditions Will Require Broad Skills Away
from Centers of Power

Prior long-arm-of-impact technologies allowed organizations to respond to their environment. A long-established idea in business is that an organization's environment must determine how it functions. In organizational design, stable markets and long–life cycle products allow choices of strategies, structures, and systems that would be deadly in fast-paced environments. In leadership, a similar logic also holds:[8] A democratic leadership style may be ineffective in an environment that requires showing people the way. The causal directionality in both situations is one way—the inside must respond to the outside.

Digital technologies are far more powerful. They allow organizations to react to their internal environments by de-skilling (Principle 1) and upskilling (Principle 2) work, distributing work over time and across space (Principle 3), serving unpredicted and/or high-value emergent needs (Principle 5). They also allow organizations to affect their external environments by interacting with them on an ongoing basis (Principle 7) and by acquiring or releasing hard-to-control information (Principle 6). The outcomes of these interactions can traverse the world in milliseconds, not weeks or months or years. So they can upend environments everywhere.

In the digital epoch, organizational design and, more importantly, leadership, must be cognizant of the implications of this superfast bidirectionality. Most of all, it substantially raises the stakes for most decisions leaders make. Our collective focus on greed as the sole cause of the 2009 fiscal crisis kept us from learning this key lesson from the first, truly global, leadership failure in the digital epoch.

The Global Survey captured how the world has changed: 589 executive respondents answered five matched questions about all companies in general and their company in particular. They had to compare their current conditions with those their companies had experienced a decade ago. Regardless of where in the world they were based, the size of their companies, or the range of their operations (national, regional, global), their responses established that the business environment had become more VUCA (see figure 4.3). Fifty percent or more "Strongly Agreed" or

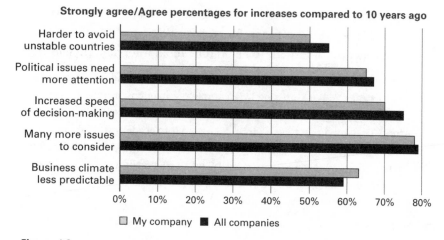

Figure 4.3

Compared to ten years ago, the business environment has become more VUCA.

"Agreed" that it was harder to avoid doing business in unstable countries, political issues needed more attention, speed of decision-making had risen, there were more issues to consider, and the overall business climate was less predictable. The single biggest change for respondents, with 78% support for their own companies, was in the number of issues they had to consider.

Away from corporate offices, on battlefields, the US Army learned a key lesson most companies still have not: VUCA environments challenge traditional leadership capabilities. In a 2010 *Harvard Business Review* article, David Weinberger described a conversation with two midcareer officers:

> [The] officers were eager to counteract…common stereotype of the paradigmatic military leader as General George S. Patton standing in front of a giant American flag. Patton was a leader when the U.S. won wars by bringing massive resources to bear on concentrated points. Iraq and Afghanistan have changed that…it's not enough to be able to do the job of the person above you. You have to be able to do 18,000 different jobs. You have to be able to manage water systems, run a town hall meeting, issue micro-grants, be politically savvy—and that's if you're a 25-year-old [sergeant].[9]

As other officers learned the same lessons, US armed forces' leadership training assimilated them. In 2011, General Martin Dempsey, the former chairman of the Joint Chiefs of Staff, that is, the senior-most officer across all branches, described the transformation discussions in progress. These would affect officers who were nowhere near battle-fields in order to train them to be on one. In an interview, he said:

> I personally believe that what made me adaptive was being pulled out of my comfort zone.... Put someone in a completely new environment, and they will have to adapt.... Should we allow officers to take a sabbatical from the Army? They might go to work in industry, go into academia.... We are doing a lot of study on defining what constitutes a broadening experience ... I am talking about it as one instrument of leader development.... If we had a menu of broadening opportunities and gave them some ability to collaborate on it, we'd be in a better position.[10]

In the opening chapter of his 2015 book, four-star US General Stanley McChrystal wrote about his time with the Joint Special Operations Task Force:

> The Task Force hadn't chosen to change; we were driven by necessity. Although lavishly resourced and exquisitely trained, we found ourselves losing to an enemy that, by traditional calculus, we should have dominated.... more than our foe, we were actually struggling to cope with an environment that was fundamentally different from anything we'd planned or trained for. The speed and interdependence of events had produced new dynamics that threatened to overwhelm the time-honored processes and culture we'd built.
> ...we realized we were not in reach of the perfect solution—none existed.... For a soldier trained at West Point as an engineer, the idea that a problem has different solution on different days was fundamentally disturbing.[11]

Search online and you will find that other top-notch armies have also discovered that VUCA conditions demand the ability to do many things, solve many different problems, each time in a new way. In more situations than not, officers with breadth are more suited to dealing with VUCA conditions than those who only do a few things very well.

Chapter 6 addresses how these lessons apply in business settings.

A Multipolar World Will Demand Increasing Collaboration

Digital technologies allow mission-critical tasks to be done in locations around the world (Principle 3). So the power to make consequential decisions is rarely concentrated in one individual or organization. More

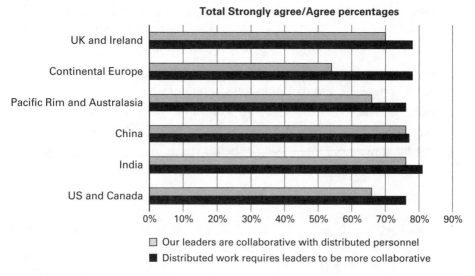

Figure 4.4
The collaboration gap: Executives fall short of expectation worldwide (for regions with $n > 50$).

than ever before, if they contribute intellect (Principles 4 and 5), decision makers have the ability to push back if they disagree. Particularly in combination with the prior two discussions of diversity and VUCA conditions, these suggest a need for effective collaboration.

Despite all the commitments to win-win in every possible instance, the Global Survey showed that business leaders don't collaborate well. Executives responded to two matched questions. One asked whether distributed work required leaders to be more collaborative. Another asked whether leaders in their own companies were collaborative with distributed personnel. The results for geographic regions with at least 50 respondents (see figure 4.4) show a gap, sometimes a very substantial one, around the world.

Continental Europeans again were either more self-aware or more self-critical. Even in regions with nonexistent (e.g., China) or small (e.g., India) differences, the intensity of support fell. Among Chinese executives, "Strongly Agree" dropped 11% and "Agree" rose the same amount, while among Indian executives, a 7% fall in "Strongly Agree" got distributed to "Agree" (2%) and "Somewhat Agree" (4%).

In the digital epoch, collaboration poses unique challenges. Jean-François Baril was chief procurement officer during Nokia's heyday. He had grown and overseen its partnership-driven procurement network, bringing together intellectual property created and owned by multiple global companies and embedding them in Nokia cell phones. I've formally interviewed him several times and informally discussed leadership with him many times more. Once, he told me at the start of a scheduled two-hour interview that he had to shorten it by fifteen minutes. He had an emergency board meeting right after ours:

> I have to report on a major problem at our partner. Battery production is tricky; fires do occur. The main factory supplying us had a fire. I talked to the top guy there and we decided to stay out of this while our people jointly figured out what to do. If we got involved, people would have focused on us, not on what they needed to do.

I expressed amazement that he hadn't rescheduled our meeting. After all, the fire threatened the supplier's ability to meet its shipment obligations. I wondered aloud how the top executives on each side could not be personally managing the crisis. He waved away my concerns:

> There's nothing for me to do right now. Their people and ours will find a solution. I'll tell the board that when we have a solution, they will know. This is what collaboration means. Selecting your partners carefully and trusting them when things go wrong.

The collaboration between the two teams could have failed for several reasons. First, the people involved worked far apart and, unlike colocated people, didn't know each other well. Digital technologies mediated much of their interactions. It's harder to build trust and collaboration at a distance than when people are colocated.[12] Second, because of cultural and ethnic differences, they communicated and made decisions very differently. Nordic Europeans can be very direct, while some other cultures are often guarded. These conditions can sometimes challenge trust and collaboration.[13]

Third, both sides had to involve people with critical capabilities and knowledge who hadn't worked together in less demanding, business-as-usual conditions. Building such "risky trust"[14] is difficult. Those who regularly rely on it usually work face-to-face, as in hospital operating theaters, where institutional guardrails exist. Distance makes this more challenging.

Fourth, the possibility of huge losses and lawsuits at stake at the corporate level and heightened stress at the team level could have triggered interpersonal conflict. That makes it harder to talk about, much less resolve, work-related conflict.[15] Finally, solving the problem required significant creativity under conditions where few good options existed.

The teams at Nokia and its partner performed exceptionally well. Outsiders didn't notice the crisis, which had jeopardized millions of cell-phone sales globally.

Chapter 7 addresses what it takes to go beyond the banality of win-win.

Creativity and Synthesis Will Be More Important Than Productivity and Analysis

In the twentieth century, it was easy to separate creation (primarily R & D- and marketing-related activities) from exploitation (all else). In contrast, digital technologies are increasing the importance of cerebral work (Principle 4). "Good work" is taking the form of ideas, concepts, and symbols to be manipulated on computer screens—across all areas of organizations. New ideas, events, projects, customers, and markets must regularly be created (Principle 5)—across all areas of organizations. Moreover, all of this is happening in the context of distributed work (Principle 3).

As a result, the distinction between creation and exploitation is increasingly hard to maintain. My research on using knowledge to compete had led me to conclude that "every act of management must be an act of learning."[16] More recently, Amy Edmondson has written about "execution as learning."[17] Leading for creativity is no longer just a nice-to-have virtue in R & D departments but is essential almost everywhere.

Around the world, we are largely unprepared for this challenge. Primary education in most places relies on rote learning. College education often relies only on solitary capstone courses; in effect, it assumes that students will figure out for themselves how to create whole cloth out of individual spools of thread. Business processes and organization

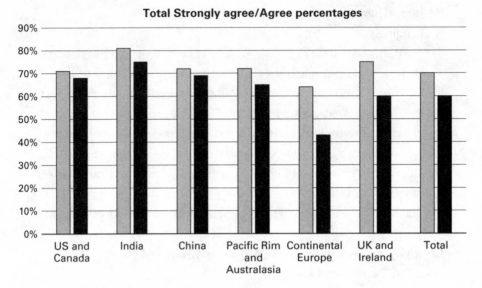

Total Strongly agree/Agree percentages

☐ In general, digital technologies have increased the thought content of work
■ In our company, we have to think more than we had to

Figure 4.5
The thinking gap: In general, thought content of work has risen, but executives themselves don't have to think more (for regions with $n > 50$).

structures still embody the percepts of scientific management and the quality movement and use temporary, superimposed artifacts to synthesize.

Executives who responded to the Global Survey revealed two particularly telling gaps. The first showed up in their responses to two matched questions about whether digital technologies sharply increase the thought content of work in general and whether they had to think more than in the past. The gap for "Strongly Agree" and "Agree," averaged across all respondents, was 10% (see figure 4.5). While the responses were high and well-matched for the United States and Canada, India, and China, they weren't in Continental Europe (21%) and the UK and Ireland (15%).

One plausible explanation, of course, is that Continental Europeans are truly pessimistic. This time, however, they have company. Another plausible explanation is that work has been redesigned in some parts of the world; areas that haven't done so have to catch up. The most likely

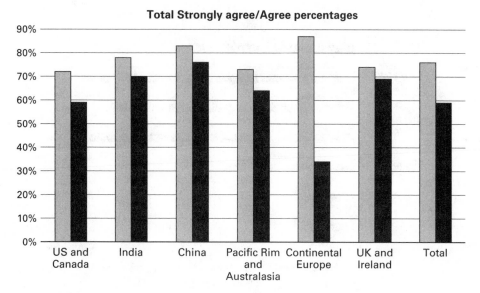

Figure 4.6
The creativity gap: Though there is a need to foster creativity and promote learning, executives don't think their own leaders do so well (for regions with *n* > 50).

explanation, however, is the simplest explanation: People tend to be optimistic and believe that difficulties others will endure will bypass them. So they assume they don't have to change.

The Global Survey also showed that respondents recognize the need to foster creativity and promote learning. They also believe their organizations are failing (see figure 4.6). Here, no region of the world can claim success. On the basis of combined "Strongly Agree" and "Agree" responses, around the world, they rated the need for fostering creativity and promoting learning higher than they rated their own executives' ability to do so. The average shortfall was 17%; the gap was 13% in the United States and Canada and a whopping 53% in Continental Europe. Areas that seemingly did well did so on the strength of their "Agree" responses. These exceeded their "Strongly Agree" responses: India (6%), China (14%), and the UK and Ireland (11%).

Chapter 8 addresses how leaders can make creativity a primary focal point of leadership.

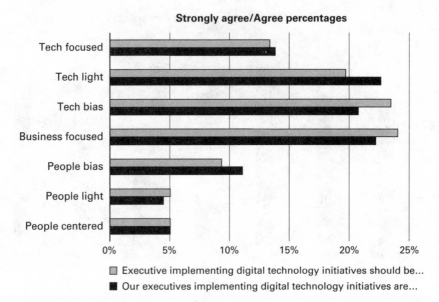

Figure 4.7
Executives worldwide aren't taking people's needs into account while implementing digital technology projects.

The Focus of Technology Initiatives Will Need Rebalancing

The Global Survey asked respondents two matching questions on digital technology implementation: Who should drive it and who was driving it in their companies? It offered seven possible responses ranging from "Technology focused" to "People focused" with "Business focused" at the midpoint. Fifty-six percent of the respondents collectively picked the three technology-skewed choices for the "should" question, and 58% collectively picked them to characterize their corporate realties (see figure 4.7). "Business focused" attracted only 24% and 22%, respectively, while 19% and 20%, respectively, picked the three people-skewed options.

We have ample evidence from our knowledge of design and creativity that technology and/or business focus, without attention to people's needs, produces suboptimal outcomes (see chapter 8). So these results are an early warning signal. Another problem is more acute. Earlier, the chapter 2 argument against the abuse of the term "disruption" said,

"Words matter. They shape our thoughts, just as much as our thoughts shape them." Similarly, unbalanced pursuits of the next new shiny bauble can easily lead to unwanted or socially harmful or even dangerous outcomes.

These issues, and what leaders need to do to avoid getting entrapped, are discussed in chapters 9 and 10.

People Are Gaining Power over Key Decisions Sooner

The final implication, applicable across all subsequent chapters, is attributable to the distribution of work (Principle 3), aided by upskilling (Principle 2). People are gaining the authority to make consequential decisions far sooner than they used to in the predigital world. So they don't have the lengths of time to prepare that they previously did. "Time to expertise" is being severely compressed in the early twenty-first century.

Between 2005 and 2010, Michael Watkins and I informally recorded data from midtier executives around the world who attended the leadership programs we taught. One question asked the number of months each had been in the most recent position held. The response never fell below twenty-four months, and unless the executive was a scientist or technologist, never exceeded thirty-six months. So in two to three years, the average executive had to transition into a job, master its idiosyncrasies, deliver noticeable value, find another job, and move on.

The Global Survey formally pursued a related issue. It asked for the number of years of experience the respondents' companies required now, and had required ten years ago, before giving people cross-organizational or cross-border responsibilities. Six hundred executives answered both questions (see figure 4.8), making their responses eligible for analysis. Though 199 companies—a third of the respondents—didn't change their policies, overall, companies requiring over ten years of experience dropped about 8%, while those requiring one to five years rose almost 6%. China bucked this trend; almost 64% of its companies required the same or more experience.

The data for multinationals is even more compelling: 288 out of 430 companies (67%) actually changed their requirements. Among them "1–5 years" rose 12.5%, while "over 10 years" and "6–10 years" dropped 10.8%

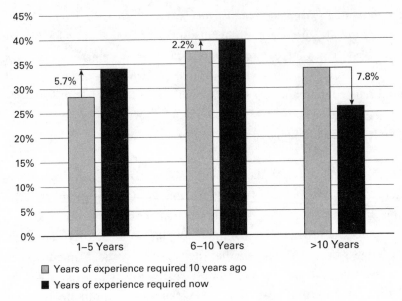

Figure 4.8
Years of experience needed for cross-border or cross-organizational responsibilities (data are for 600 survey respondents).

and 1.7%, respectively. Consequently, across all 430 multinationals, people with a lot less experience were making these consequential decisions.

These findings establish that in the digital epoch, people are affecting the lives of many others in their organizations and outside them relatively early in their careers. Organizations need to pay attention to individuals' leadership abilities far sooner than they used to. Leadership is no longer just the responsibility of those who have reached the highest echelons.

Jaikumar's entire research can be reduced to one simple idea that the prior two chapters and this one have captured: Ideas, processes, systems, values, and organizations that work beautifully in one epoch fail to be sufficient in the next. The same is true for leadership.

You can't afford to ignore this lesson. "Do-overs" may be hard to come by. The rest of the book will address what you must do.

II Leading for Creativity

Others have seen what is and asked why. I have seen what could be and asked why not.

—Pablo Picasso, *Pablo Picasso: Metamorphoses of the Human Form: Graphic Works*

When I was a kid, there was no collaboration; it's you with a camera bossing your friends around. But as an adult, filmmaking is all about appreciating the talents of the people you surround yourself with and knowing you could never have made any of these films by yourself.

—Steven Spielberg, Academy Award–winning director, producer

5 The Many Faces of Talent

Sam Palmisano became IBM's CEO in March 2002. His predecessor, Lou Gerstner, had successfully fended off Wall Street's efforts to dismember IBM and had returned it to health. So Palmisano could focus on reinventing it for the digital world.

At a 2005 analyst meeting, he described IBM as a "globally integrated enterprise (GIE)," not just a "multinational." In a 2006 *Foreign Affairs* magazine article,[1] he essentially defined this to mean the ability to effectively work across time and geography as envisioned by Principle 3.

In Palmisano's GIE, work, including mission-critical research, crossed corporate and national boundaries, happening where it could best be accomplished. (IBM currently conducts research on cognition, key to its future, in six very different countries—America, China, India, Ireland, Israel, and Switzerland.) This sometimes required collaboration with competitors on specific projects, as has happened with Microsoft. It could also require cooperation among citizens of countries whose governments had acrimonious relations.

IBM had been preparing its executives for these changes since 1995. Gerstner had reversed a decades-long policy: Instead of "being blind to differences" among peoples, the new approach drew "attention to differences, with the hope of learning from them."[2]

Palmisano established a series of high-level "Integration and Values Teams" (IVTs).[3] The first three of these globally integrated IBM's supply network, manufacturing, and human capital, respectively; the fourth created global solution centers.

The fifth IVT built on the foundational work on people initiated under Gerstner. Led by two of Palmisano's direct reports and staffed by fifty senior executives from businesses, functions, and key markets worldwide, it took on the transformation's most challenging element. It aimed to define "the Global IBMer"[4, 5] and create a culture that would produce many global leaders, not just Westerners or West-based foreigners. It drew on the outputs of a twenty-four-hour online "ValuesJam" held in 2003. Three hundred thousand employees globally had debated possible changes in IBM's long-standing values to make them relevant worldwide. Since then, IBM had added ninety thousand employees in emerging markets. What, the IVT5 debated, would ensure that IBM employees were unified in their diversity?

Other leading global companies are also focused on attracting diverse leaders. Rio Tinto, a Global 200 company that operates in countries that range from democracies to dictatorships, employs highly educated professionals and others who live in parts of the world where employment opportunities are scarce. In 2016, under a new CEO, it began restructuring how it worked. The effort included reassessing leadership capabilities of aspirants to top positions. Referring to a recent CXO search she oversaw, Vera Kirikova, the chief human resources officer, said:

> Interest in the human being.... Not as a means to an end, but actual respect for the individual. We need to make our company people-centric. People are not the means to a strategy.... [The candidate must understand] what we do, and how that affects people.
>
> We used to recruit only based on competencies. Now we look for cultural fit, values fit.... We want to take exceptional talent from every nationality.

From well before the digital epoch, Unilever had periodically moved high performers into unfamiliar territories and roles. A promotion didn't necessarily accompany these "learning and proving" moves; they exposed executives to the cultural diversity of the people they were meant to lead and the markets they had to serve.[6] Despite this extensive cross-pollination, by 2015, promotion rates of Asian executives[7]— except for Indians like its then chief operating officer (COO), Harish Manwani—lagged those of their Western peers. This problem extends well beyond Unilever.[8] An article coauthored by Winter Nie, Jean-Louis Barsoux, and a senior Unilever Asia HR executive, Daphne Xiao, generalized the problem:

[M]any Western MNCs [multinational corporations] rely on a set of global leadership competencies. In many cases, the requirements they use to select global leaders are essentially the same as those used to select domestic leaders, making it extremely difficult for Asian managers to compete. The reality is that corporations can't turn Easterners into Westerners or vice versa.[9]

The article recommended "reducing the number of must haves," "reviewing HR practices…to identify outcome biases, blind spots, and gaps," and seeking "East Asian voices and business perspectives at the table." In addition to working on these issues internally, Unilever approached IMD to custom design executive workshops for global companies. Senior Western and Asian executives would work on problematic practices and biases in their companies, seeking inspiration and ideas from each other. Since bias isn't just a Western problem, Asian-headquartered global companies were also invited to participate.

In 2013, Johnson & Johnson, whose human resource policies had for decades been a benchmark for others, began a three-year effort to adopt culturally neutral, globally coherent leadership standards.[10] These now cover all employees except where prohibited by law or preempted by contracts.

Daimler's top leadership initiated its "Leadership 2020" effort in 2015. Eight "design teams" located across the world, each with a mix of senior and midtier executives, began formulating global standards of leadership to replace existing ones.[11] (Disclosure: I delivered a keynote address for the Pacific Asian team, which was recorded for the benefit of the others.)

In a 2015 interview with a journalist, Salesforce CEO Marc Benioff expressed regret that he had not paid attention to inclusion when he founded his company. He said:

It's part of our training [that CEOs] are supposed to focus on the shareholder [earnings per share]. [Economics Nobel laureate] Milton Friedman said it best, "The business of business is business." Well, guess what, it's 2015 and the business of business is *not* business. The business of business is improving the state of the world. And that's what CEOs need to do. They need to build great products, take care of their employees, customers. They need to build great companies with great cultures. But they need to make the world a little better too.[12]

Before the interview, Benioff had taken a very public position supporting LGBTQ rights in a US state that passed a discriminatory law. He
had urged other CEOs to do the same. During the interview, he said that
though the percent of women in his company was less than 30%, he
wanted to increase it to 50% in five years. As a result of an audit Salesforce
had conducted, it had found imbalances in the salaries paid to men and
women. It responded by increasing salaries for 6% of the staff. It had man-
dated changes in hiring and promotion policies so women and minorities would always be included in applicant pools. Another policy required
women to constitute at least 30% of the attendees in any meeting.

The Digital Epoch's Case for Inclusion: It's Existential

Whatever else they may be, top executives of leading global companies
are usually pragmatists, not ideologues. Instead of chasing abstract goals,
they strive to make their companies successful by growing revenues and
profits, ensuring cash flows, launching new products, and entering new
markets. Why, then, are they investing in redefining leadership and
changing talent management policies to become more inclusive?

Clearly, there has been legal pressure mandating inclusion, primarily
from Europe. Additionally, a drumbeat of research by academics and
consultants has positively linked diversity to corporate performance on
the metrics used to measure CEO performance.

For example, the presence of women on boards of directors has been
found to improve firm performance in America,[13] Spain,[14] and The
Netherlands[15] (but not Australia[16]). Their presence also raises corporate
social responsibility ratings and has a positive impact on reputation.[17]
Cultural diversity improves decision-making even when teams are dispersed.[18] A 2017 McKinsey study that looked across one thousand companies in twelve countries—and within them, at the total workforce,
executive team, and board of directors levels—found that the most
gender diverse companies performed substantially better on metrics
relevant to financial markets than the least diverse ones. The study also
found similar results for ethnic and cultural diversity.[19]

Two digital-epoch Principles are also making the need for rapid change inescapable. Work is becoming more cerebral (Principle 4) and is being done across geography (Principle 3) in multicultural settings by men and women. Inclusionary mindsets, behaviors, and actions are essential because limiting one's access to talent is unfathomably illogical. The history of long-arc-of-impact technologies suggests that mastering digital technologies won't determine winners; making the necessary organizational and leadership changes will.

It behooves executives to remember that Japanese companies, which developed the core ideas of teams and inclusionary leadership during the quality movement, had a thirty-year lead on the rest of the world. However, they failed to extend those once-revolutionary ideas to attract and retain talented foreigners or, indeed, Japanese women. These factors contributed to their loss of once-dominant positions.

A 2011 *McKinsey Quarterly* article about Japan's travails made this point.[20] It urged Japanese companies to adopt English as their working language because "it opens up a world of talent.... Talented Asians with global aspirations tend to learn English, not Japanese or Russian." It also recommended that they "[e]mbrace diversity and set aspirational targets for women, foreigners, and Japanese managers from other companies and industries."

So the issues discussed hereon are universal, though I exclusively present Western examples. This bias in the selection resulted solely from my perusing works published in English. Moreover, it's important to note upfront that discussions of one type of inclusion (say, gender) could easily be replaced by discussions of other types (say, racial or cultural) that challenge societies worldwide.

Developing the Abilities to Be Inclusive

You must first accept that *people's identities matter*. People take pride in their respective genders, heritage, achievements, personalities, and a host of other related characteristics. Their perception of assumptions that others make about them can profoundly affect their ability to contribute effectively. Assumptions driven by societal, institutional, and individual biases are often grossly flawed.

In 2017, after sustained protests led by women—including a walkout by twenty thousand employees globally—Google changed policies that shielded it from sexual harassment, discrimination, and wrongful termination lawsuits.[21] One male software developer who believed "Women can't code" triggered the crisis that led to the walkout.[22] Consistent with Google's practices, he posted a "manifesto" on Google's intranet, essentially claiming that Google's efforts to be inclusive were inhibiting his own advancement. Google's CEO, Sundar Pichai, fired him when the public disclosure of the manifesto (the Principle of radical transparency at work) led to widespread criticism of Silicon Valley's bro culture.

The coder made a major identity-related assumption: "Women can't code." This belief arose from his experiences in college. His compatriots in coding classes had largely been male. His belief ignored the history of coding in the United States.

In the mid-twentieth century, when US women weren't allowed to be engineers, scientists, and professors of mathematics, many became the unrecognized "computers" who did the calculations and the programming that expanded our knowledge of astronomy and also put men (of course!) into space.[23] As societal mores changed, women joined companies and universities.

In 1960, when women made up only about 36% of the workforce, they accounted for 27% of the "computing and mathematics profession" (the official US government classification).[24, 25] In 1990, women made up roughly 46% of the workforce and 35% of the profession. By 2013, when women made up 47% of the workforce, their representation in the computing and mathematics profession had dropped back to 26%.

Data on women receiving undergraduate computer and information science degrees follow a similar pattern. In the 1970–1971 academic year, 13.6% were women. This percentage rose steadily till 1983–1984, when 37.1% were women. It dropped almost every year subsequently, and in 2010–2011, only 17.6% were women.

Jane Margolis found these patterns resulted from academia's magnifying of preexisting social biases.[26] Personal computers had just become available during the 1983–1984 academic year. American families often put the first one they bought in their sons' rooms. Fathers worked with sons on coding; no one worked with daughters. By the time the sons and daughters arrived in college, the sons knew the coding taught in the introductory classes while the daughters did not. Universities changed curricula to accommodate the sons and left the daughters behind. In 2002, Margolis published her findings in a book coauthored by computer scientist Allan Fisher;[27] he had originally asked her to do the research. Universities are now changing how they teach introductory computer science. Those that have done so are seeing more women opt for that major.

In contrast, in Latin America and India, women account for higher proportions of computer science students.[28, 29] In India, where social biases are different, it's currently 40%.[30] Parents encourage girls to learn coding because they consider the work environment for coders and computer scientists safer than for many other professions. *So biology isn't, and never should have been, destiny.*

Yonatan Zunger, a senior Google engineer who left to join a start-up shortly before this incident, blogged:

> All of these traits which the manifesto described as "female" are the core traits which make someone successful at engineering. Anyone can learn how to write code;…The truly hard parts about this job are knowing *which* code to write, building the clear plan of what has to be done…and building the consensus required to make that happen.
>
> …the conclusions of this manifesto are precisely backwards. It's true that women are socialized to be better at paying attention to people's emotional needs and so on—this is something that makes them *better* engineers, not worse ones. It's a skillset that I did not start out with, and have had to learn through years upon years of grueling work [emphases in original].[31]

The male coder at Google was flat-out wrong. We can thank him, however, for teaching one key lesson. Many articles and websites offer advice on leading noncolocated, multicultural, multi-organizational teams.[32, 33] With rare exceptions,[34] they prescribe "five rules which immediately improve your leadership" of such teams. Those rules actually won't help you *unless you believe that people's identity matters* and must be respected. You can't do the right thing consistently unless you actually care it is the right thing to do. *No set of rules would have prepared the coder to lead a team that included accomplished women.*

Research done at a country/industry level supports this assertion: "Countries and industries that view gender diversity as important capture benefits from it. Those that don't, don't."[35]

You also have to accept that *behavioral diversity*—differences in accepted cultural practices, behavior, or language—*matters*. It is *not* "best regarded as a necessary evil: something that no global business team can avoid but the effects of which the team must attempt to minimize through language training and cultural sensitization."[36] This perspective, if ascribed to, will compromise your ability to lead.

Michele Gelfand and over forty colleagues surveyed 6,823 people in thirty-three countries.[37] They defined fifteen situations from various aspects of personal and professional life and asked about the appropriateness of twelve behaviors in those situations. They also assessed many ecological and historical factors in each country. They concluded that people's behaviors have evolved in response to their environmental conditions. Those who come from societies exposed to accentuated threats tend to stick close to tightly delineated social norms (e.g., be on time). Those who come from more relaxed environments have more forgiving behaviors (e.g., be more open to change and new ideas). They even observed these patterns across the fifty states in the United States.[38]

Behaviors that are engrained over generations can be hard to modify. As work becomes more thought driven (Principle 4), restricting how people naturally behave *without good reason*, or requiring them to adopt behaviors that are unnatural for them, can inhibit their ability to contribute effectively. These actions can also produce misunderstandings and conflict.

You must be aware of possibilities, pick up cues, and make allowances for differences in behavior. Notice I'm urging you to "make allowances" and *not* "be clear about expected behavior." The latter leads to useless advice such as "Have everyone sign off on a team charter that prescribes expectations." Such advice seeks to formalize behaviors that particular cultures consider virtuous. They work only as long as all involved buy into those virtues. When they don't, all the charters, service-level agreements, and legally binding contracts in the world merely become the means to attribute blame and assign cost for failure.

In addition to anticipating possibilities and making allowances, be ready to adapt. You have to choose from a range of acceptable options to achieve the desired outcome; you may not even recognize the need for options if you lack empathy for behavioral diversity.

How to Learn What You Need to Know about Identity and Behavioral Diversity

Learning to deal with behavioral diversity basically comes down to exposure to differences.[39] It requires people to evolve from ethnocentricity (which can range from denial to denigration of other cultures) to ethnorelativity (which seeks out differences, accepting their importance and adapting) to integration (which is having a multicultural worldview). For example, Gerstner's changes at IBM formalized ethnorelativity, while Palmisano was pursuing integration. These changes don't happen overnight, but they won't happen ever without effort.

Wherever and whenever possible, ask people about diverse behaviors and identities in an unthreatening way ("I noticed…" "Did I get that wrong?" "Help me understand…"). Asking such questions can be tricky, given the sometimes fraught dynamic of people in privileged positions. Recognize that it's *never* the responsibility of those who aren't privileged to explain to those who are why a particular statement or practice or policy is problematic. However, if you treat them with respect and if such conversations lead to constructive changes, they will help you understand what is important to them and why. Psychological safety, a concept discussed at length in chapter 7, is critically important for doing this effectively. In psychologically safe environments, people are protected and supported when they speak up.

An Inclusive Leader at Work

To effectively lead diverse peoples, first **eliminate the "language of leadership" problem from your vocabulary.** In 2008, long before it changed its global leadership standards, Johnson & Johnson's annual

360° reviews for executives included two career derailers: "Indecisive—shows reluctance to commit to decisions" and "Avoids needed conflict."[40] The first, given that decisiveness is a quintessentially Anglo-Saxon male attribute, discounted the deliberative decision-making practiced in Pacific Asia and by many women.[41, 42] The second implicitly assumed that conflict was sometimes good. But for whom? The two standards existed to ensure that people in leadership positions could make tough decisions. Why not simply say that in inclusive ways?

Even well-intentioned, careful companies and individuals struggle with the language of leadership, so deeply is it ingrained in our cultures. Professors and consultants who routinely caution executives to eschew sports analogies or idiomatic language in cross-cultural settings don't urge them to monitor their leadership-related language.

You won't catch every "language of leadership" term you need to change. Focus on those you use commonly. Ask others who are different from you what these mean to them, while being cognizant of the challenges of having these fraught conversations. When you learn something, make it a point to apply it. Great companies and real leaders make necessary changes even if doing so is difficult.

Second, **address perceptions of unfairness wherever possible.** Think about a recent meeting in which many people were colocated with the boss, but key members were disembodied voices on a speakerphone or small images on a screen. Was the meeting scheduled for the convenience of the colocated people, or did it consider those elsewhere? Could those who were afar socialize and build bonds? Could they get their voices heard? What happened if a woman made an unexpected point? Or if a man made the exact same point soon after? How did people with weak command of the common language fare? Who got the choice assignments?

We all have experienced what often happens. The meeting is scheduled for the convenience of the boss. It moves from agenda item to agenda item, supposedly to be respectful to those far away but actually depriving them of the ability to socialize with peers. The boss asks questions or seeks comments from those around the table and overlooks those off-site. Women get ignored, or men step in to explain "what she was trying to say." Men who make identical points as women are commended for "out-of-the-box thinking," while the women aren't. Those who don't speak the common language well usually don't speak, or if they do, get interrupted when they struggle for a word. People onsite get choice assignments; those off-site may be asked to pitch in. Men get choice assignments; the women are told to help, even if they really deserve the project.

Over time, resentments and frustrations build up. Websites like Glassdoor promote radical transparency (Principle 6). It's much harder today to hide providing unequal compensation for comparable work or giving some people access to resources, power, and opportunity while denying the same to others. Consequently, the best of those who feel slighted seek other opportunities.

Fix what you can to assure "outcome justice." IBM's Palmisano traveled the globe extensively to make himself accessible to those far from Armonk, New York. He distributed decision rights—who has the right to make which decision when?—to executives far away; this simple step balances out the normal jockeying for power that otherwise favors those who are colocated with the boss.

Former ABB CEO Goran Lindahl's tweak of a predecessor's policy is particularly relevant to the digital epoch. He famously declared that the official language of ABB was not just English, but "poor English."[43] That beautifully crafted policy made it easier for everyone to speak.

Not all outcomes can be made fair: While salary differences among people doing similar jobs at the same location are problematic, differences across geographical locations may be rooted in legitimate economic differences. Take people's inputs into consideration. Ensure that the decision-making is unbiased and transparent. Explain what you can't do and why. Assuring such "procedural justice" can be powerful: Reasonable people accept solutions they dislike if they feel the process was fair.[44]

Third, **reduce the mutual incomplete knowledge problem.** Simultaneous with distributing work (Principle 3), digital technologies reduce opportunities for face-to-face contact. Noncolocated people working together most likely haven't "broken bread" with those whom they must trust. They may make assumptions about other people's environments based on their own. They set up web-based meetings assuming everyone has the necessary bandwidth. They assign five minutes for bathroom breaks without knowing where the bathrooms are at other locations. A former senior Indian executive of a company that worked globally provided a compelling example:

> I can recall countless instances where Westerners showed no respect for our festivals to schedule visits and reviews, project deliveries and support calls. I have been on conference calls on Diwali just as we started [prayer ceremonies]....The assumption was that the Indians would work their butt off no matter what. But when it came time to schedule anything at our convenience we would be told it's Christmas or New Year's vacation, or family time or whatever. This absence of respect...led to purely contract-driven relationships with Westerners, with almost no emotional commitment to the outcome.
>
> However, when we worked with other Asians, mutual respect...led to much smoother [interactions]. Calendars...were respectfully blacked out in advance and mutually agreed to, including for personal family events. Huge difference in team dynamics....Huge difference in behavior when crises happened.

A case study about the erstwhile computer company Sun Microsystems, meant to teach how to repair fraught cross-boundary interpersonal relations, offers another telling example.[45] Its protagonist, a "high potential" young American manager, had flown to India and quickly recruited a large team of engineers who were graduates of the famous Indian Institutes of Technology (IIT). Their mission-critical work? Routine, but urgent, essential maintenance of existing systems that served global clients. Months later, their discontent had devolved into acrimonious finger-pointing with their colleagues in America and elsewhere.

Seasoned executives conclude that the team wouldn't last long: Many IIT graduates have become billionaires in Silicon Valley, while others work in start-ups with billion-dollar capitalizations in the Indian versions of the Valley. They predict that frustrated by the mismatch between their abilities and their assignment, the Indian team members would leave en masse, possibly to join a start-up.

We may not know bandwidths, or where the bathrooms are, or when major festivals are celebrated. But these gaps are not hard to fill with a little effort. The IT department can check bandwidth issues. Digital calendars can highlight major religious events elsewhere. A few minutes online could have told the young American manager about the IITs. *The real issue, bluntly stated, is: Do you care?* To be an effective leader in the digital epoch, you must.

Members of your team must also ameliorate their mutual knowledge problems. Create environments in which they can discuss their differences and find solutions. Tie these conversations to your firm's or team's values: They don't need to become best friends but must build relationships that enable them to collaborate and cocreate. Find opportunities for people to socialize; if budgets preclude even occasional face-to-face meetings, introduce buffer time into conference calls. Encourage peer-to-peer communications to supplement, or even replace if appropriate, formal decision-making mechanisms. If budgets allow, visit each other's turfs and shadow your counterparts. If you control the budget, earmark funds for such mission-critical contact.

Fourth, **seek inputs from people with different perspectives.** Our brains filter our environmental stimuli. We prefer the familiar to the unfamiliar.[46] We recognize friends in crowds but are terrible eyewitnesses of events, particularly those involving strangers. We see what we believe (you read that correctly).[47] We value our own conclusions more highly than those of others. We judge without understanding context.

Diversity helps us compensate for these biases. It improves the organizational climate for innovation and lowers turnover.[48] Educational and age diversity improve team performance when members have a high need to engage in cerebral work.[49] Given difficult problems to solve, teams in which members had different approaches to encoding the problems and searching for solutions outperformed the best individual performers. In an academic article about his book on the subject, Scott Page called this "[t]he diversity trumps ability theorem."[50]

Fifth, **understand the power of implicit bias and take mitigating actions.** Implicit bias refers to "relatively unconscious and relatively automatic features of prejudiced judgment and social behavior."[51] In effect, it is social bias—identity and behavioral—that is so deep-seated that people are not aware of their biased behavior, which may violate their espoused beliefs.

For example, students at a top US medical school favored Caucasian and upper-class patients relative to those who weren't in these groups.[52] Venture capitalists—both men and women—asked male entrepreneurs questions about maximizing gains and female entrepreneurs questions about minimizing loss; they then gave more money to the men.[53] In experiments, male professionals who expressed anger during job interviews were given higher status than women who did so; the men were considered more competent and worthy of significantly higher salaries.[54] In a widely accepted test designed to identify implicit bias (the Implicit Association Test), "ethnically White names and leadership roles (e.g., manager...) or leadership traits (e.g., decisiveness...) were paired" faster by "both White-majority *and ethnic minority* participants" (emphasis added).[55] *Both groups* were more likely to promote people with "ethnically White" names.

Implicit bias is hard to fight because it is unconscious. If you are aware of it, you may be able to reduce the possibility of its occurring. (You can test yourself at https://implicit.harvard.edu/implicit/takeatest .html.) Talking about possible solutions with those who experience such bias can help. Those conversations, which may be very difficult, should only take place in environments of heightened psychological safety.

You should also look for policies and processes you can modify to reduce the possibility of implicit bias. For example, in recent years, several companies have begun removing all identifying information (names, addresses, universities attended, etc.) from incoming résumés before any potential decision maker sees them.[56] Doing so focuses the decision maker's attention on the candidates' competencies, skills, and experiences.

Finally, **tackle the difficult task of assuring consistency across individual acts of inclusion.** At IBM, Palmisano assigned two very busy executive officers to colead the effort to define "the Global IBMer." That's because developing globally coherent principles that would meet the expectations of several hundred thousand people was a fiendishly difficult challenge. His choice of executive officers—the chief technology officer and the chief human resources officer—also signaled that inclusivity was a mission-critical issue for technology development.

You will also face similar challenges. At some point, you will have to weave your ad hoc responses to modifying your language of leadership, or assuring outcome and/or procedural fairness, or improving mutual knowledge, into coherent standards and policies. If not, individually correct but mutually inconsistent decisions could collectively produce undesired outcomes.

Two simple questions can help. First, explore the shortcomings of your current standards of leadership by asking "How can this standard/policy hurt us in a digital world?" This will provide you with a clear rationale for change. Next, assess the possible new leadership standards or policies by asking "How will making this change bind us together in a digital world?" This question will expose possible flaws in your ad hoc solutions and ensure they are consistent with the values you consider inviolable.

In the past, personal and organizational ethics—and sometimes, laws— guided inclusive behaviors. In the digital epoch, while these continue to be important, cerebral, distributed work is making inclusivity an existential imperative. No matter where in the world they work, leaders in the digital epoch must behave accordingly.

6   A Broad Wingspan, Not a Long Tail

For most of the twentieth century, companies worldwide promoted "I-shaped" executives to leadership positions. They had deep expertise in a single area—a function or product line or technology or business. Each promotion broadened their responsibilities slightly. The process worked in the scientific management epoch because many rungs led to the top. During the quality movement epoch, cross-functional teams compensated for the reduced number of rungs.

In the digital epoch, executives lack the time to learn incrementally. As discussed in chapter 4, the time needed to reach positions of consequential authority has shortened. Speaking of midtier executives with I-type backgrounds, the chief technology officer of a top global consumer goods company summarized the challenge:

> If I step out of my office and tell my people, "Go take *that* hill!" it will be done. No question. On time. On budget. Brilliantly. If I step out and say, "Go take *a* hill!" they will freeze. "Which hill?" they will ask. The trouble is, we live in a world in which increasingly, they must regularly decide which hill merits taking.

In a 2010 interview, design firm IDEO's CEO Tim Brown nudged businesses toward a possible solution.[1] *All* IDEO professionals had expertise in one area but could empathize with the perspectives of other disciplines and even work with (some of) their tools. This "T-shaped" broadening of knowledge and skills helped them create their many first-ever-in-the-world designs with very little top-down direction. Brown noted that other design powerhouses (Apple, P&G, and Nike) also hired such professionals.

Almost twenty years ago, the T-shape concept was used to describe business-savvy IT experts (but not IT-savvy business experts!). That acknowledged, IDEO's iconic status brought Brown's words widespread attention. Since then other top firms have championed T-shaped professionals although, unlike IDEO, they don't exclude every I-shaped person.

Marc Vollenweider cofounded Evalueserve, a digital native provider of data services to Global 500 companies, almost two decades ago. After being CEO, and then chairman, he recently relinquished operational responsibilities and became chief strategy officer. He argued for a more aggressive change, in line with those made in the US Army (see chapter 4):

> I think in terms of feedback loops and second and third order non-linear effects....(The business world has become) a non-linear coupled system...which can destabilize in a jiffy....Everyone has the same information at the same time....People are making decisions faster...signal to noise ratios are decreasing....
>
> Our profits depend on China's copper consumption because Chinese copper consumption drives the Chilean economy. When demand for copper rises, our staffing costs rise in Chile and our profits fall....So, noise and extraneous factors can be far more important than manageable change. Unpredictability is increasing and is becoming inherently uninfluenceable.
>
> Leaders have to cope with this environment. They have to envision future products that are radically different....He's bound to be wrong [a lot]....But he must proceed only knowing if we don't do it, we'll be dead....He must be in the market to get the "texture"...guys in the back office can't do [it]...true digital leaders don't delegate this communication....
>
> He doesn't want to build an empire—doesn't want that baggage. Instead, he relies heavily on external partners to help get it done....
>
> He is not a techie, but is also not afraid of getting into technical details....When faced with 24,000 different small decisions to be made efficiently, he empowers with absolutely no need to escalate.
>
> He must have the willingness to jump into stuff that he hasn't seen before...and learn the hard way.
>
> Breadth is key. By definition, you'll get into new fields. You need drill down capability not as a specialist, but you must have read all the literature.

I call leaders who conform to Vollenweider's profile "T-prime" leaders. Their abilities go beyond those of Tim Brown's design professional. They have wide wingspans (broad knowledge) instead of long tails (depth of knowledge). Their broad knowledge allows them to figure out which hills merit capture. Their willingness to learn compensates for their lack of depth of knowledge and helps them determine how to do so.

Developing T-Prime Capabilities

Years ago, leadership guru Warren Bennis, and Robert Thomas, wrote about "crucibles of leadership."[2] These relatively rare, extraordinarily tough events—launching a service or entering a major new market or shutting down a factory—taught top executives how to truly lead.

In Vollenweider's view, executives leading in the digital, VUCA world face crucible-like conditions regularly. One of his executives identified the need for a novel digital platform during routine client interactions. He then personally led the development of the platform, which quickly became a standard for investment bankers. In this one project, he stayed close to potential clients, so he had the information he needed to define his strategic intent (defined below). He partnered with other companies; their staff augmented his. He planned for uncertainty because needs kept changing. He delegated key decisions to a distributed team—and trusted them. He learned about new digital technologies and design methods (Design Thinking, discussed later in this book, in particular) and became an expert in how investment bankers work daily.

Wouter Van Wersch, CEO of GE Asia-Pacific (itself a multibillion-dollar business), elegantly echoed Vollenweider's opinion: T-Prime leaders have *"the ability to navigate the in-between places that experts avoid."* His ASEAN-based team of executives understood how ten countries—with economies that ranged from capitalist to socialist and with political ideologies that spanned freewheeling democracies to an absolute monarchy—made major decisions. They knew the capabilities of their parent company's business units (just!) well enough to put together cross-business programs no unit could on its own. They spoke at least two regional languages (and English) fluently and understood the functional, professional, and business jargons their partners used. This "multilingual" capability helped them understand each group's core needs and bridge differences.

A midsenior Japanese auto company executive gave a personal example of this multilingual ability. Though a software engineer, he was leading the development of a key mechanical system for a hybrid engine. "I'm not a specialist in engine design," he said. "My specialist colleagues explain the issues and their concerns and make suggestions. My responsibility is to listen to others and decide on a direction."

Not surprisingly for the digital epoch, the path to the C-suite is changing. In 2011, Boris Groysberg and two top executives from a global recruitment firm, Kevin Kelly and Bryan MacDonald, wrote, "We're beginning to see C-level executives who have more in common with their executive peers than they do with the people in the functions they run."[3] Michael Watkins, a globally recognized expert on executive transitions, has written about the seven "seismic shifts" aspiring leaders must make. Each requires a sharp broadening of perspectives. For example, trained as "bricklayers," they must become "architects," and experienced as "warriors," they must become "diplomats."[4] Because of the increasing diversity of people on noncolocated teams and the compression of time to positions of consequential authority, you will make the "seismic shifts" earlier than your predecessors.

Unfortunately, conventional college education and initial years of work still serve twentieth century needs. Training in one field may be indispensable for some careers (think of a research scientist studying a cancer cell), but it didn't prepare you to be a T-shaped professional, let alone a T-Prime leader. At work, you proved your worth by mastering a narrow task that would have delighted Frederick Taylor.

By the time you escaped, the damage was done. You had learned "one right way" of seeing the world. You weren't prepared for new technologies appearing well before older ones matured and for the impact of China's copper consumption on the profits of an unrelated company. Allow Vollenweider's words to guide you: *Breadth is key. By definition, you'll get into new fields. You need drill down capability not as a specialist, but you must have read all the literature.*

The research of psychologist and 2002 Nobel laureate for economics, Daniel Kahneman (and his research partner, Amos Tversky) possibly suggests why this is helpful. They have called the human brain's reliance on blindingly fast decisions "System 1" thinking.[5] Acts of reason ("System 2" thinking) are slower; they require concerted effort to recall relevant facts and consider other possibilities. The brain uses System 1 most often, grabbing at easily available information, regardless of its veracity. This makes sense: In a simpler world, running away after mistaking a large rock for a skulking predator was preferable to being eaten. In today's more complicated—even VUCA—world, however, reliance on such information, called the availability heuristic, is usually problematical because the information is likely to be wrong.[6]

Breadth of knowledge turns a disadvantage into an advantage. The brain's fast reasoning is no longer blinkered by "one right way" thinking. During System 1 engagements, T-Prime executives possess more varied, yet familiar (and so, easy to recall) information to draw on instantly than do conventional executives. During System 2 engagements, they can frame issues from multiple perspectives, follow up with experts in their jargon, and better synthesize inconsistent information.

Research in other fields also suggest breadth offers key advantages. Bilingual individuals outperform monolingual individuals in tasks that require attention or inhibition or short-term memory.[7] Thinking in foreign languages reduces biases associated with decision-making by making more possibilities available for consideration.[8] Nobel Prize–winning scientists are far more likely to be accomplished at music or the arts than average scientists.[9]

Start broadening your perspectives early. If you're still in college, add courses from outside your area of specialization. Take courses in statistical thinking—not in the arcane algebra of statistics, but in the application of statistical logic to real issues. If you're monolingual, take language classes. If you're studying business, avoid classes that lecture on theory or that use case studies merely to reinforce theory or prescribe one right answer. Take classes that use cases, experiential exercises, and simulations. To get exposure to different issues in different organizations, markets, and industries, take a first job in a consulting firm.

As a working executive, adapt these approaches to your environment. Except for incontrovertible scientific facts—the universe isn't just five thousand years old, and climate change isn't a hoax—you don't have a monopoly on the truth. You can never tell where you might find the knowledge and resources you need.

So, **actively seek opportunities to develop broader perspectives.** Subject yourself to a broad spectrum of facts, ideas, models, experiences, knowledge, thinking, opinions, and feelings. Doing so is easier than you might think. Read about issues outside your field. Lunch with people from unrelated areas and consider how they frame issues. Practice making their arguments for them: "Let me see if I understood what you said."

Second, **learn to think in terms of scenarios and simulations.** Most people look back at the past and expect the future to be the same. In a VUCA world, such "linear thinking" is a foolhardy exercise. Look up the stock price charts of companies like Ericsson (1996–2002) or Bear Stearns (2001–2008). After rising meteorically for years, they plummeted because of one (Ericsson) or several rapid-fire (Bear Sterns) key events. If the stock charts of the growth years are shorn of the corporate names, executives shown them usually predict the companies will continue to do well.

Kahneman explained why: Humans are abysmally poor at thinking probabilistically. The cause-effect relations they perceive immediately after something happens—products of System 1—are usually simplistic and possibly fallacious. To think probabilistically, they have to engage System 2.

Faced with any major decision, ask yourself what assumptions you are making. For example, one assumption could be that the economic environment wouldn't change over the foreseeable future. Another could be that despite globalization, new competitors wouldn't arise from other parts of the world. Assumptions people make are rooted in technologies, markets, competitors, partners, culture, and politics. They also ignore the possibility of VUCA events.

Select from this set two assumptions that could be most problematic. These are your "critical unknowns." How would your decision be affected if the polar opposites of these critical unknowns, individually and jointly, actually held true? If your decision would change, consider two more questions: How would you know if the polar opposites actually materialized? What would you do?

This simple exercise won't lead to "paralysis by analysis." Instead, if done regularly, it will help you avoid linear thinking and increase your effectiveness in the digital, VUCA world.

Third, **volunteer to work on, or colead or lead, ill-defined projects** with major resource, political, and talent constraints; indistinct deliverables; and unclear personal incentives. Don't worry about immediate rewards; such projects are perfect "minicrucibles of leadership."

A member of GE's ASEAN team described his part-time leadership of a project that involved business units he didn't control, governments, external businesses, public and private international funding agencies, nongovernmental organizations (NGOs), and regulatory authorities. It required several years of work but had only a limited likelihood of success. Asked how he would be rewarded if the project failed, he replied that because of all he was learning, GE "would be foolish not to look after me." If it were foolish, he said, other companies would hire him instantly. Sometime later, he got two rapid promotions that vaulted him into a senior leadership role.

Being an Effective T-Prime Leader in a Digital, VUCA World

Three steps will help you handle digital, VUCA conditions effectively.

First, **ensure your leadership team has access to key knowledge bases.** Responding to the digital, VUCA world requires bridging gaps in critically needed knowledge. So ensure that people with such knowledge are at the decision-making table. Only then will you be able to navigate the in-between spaces, communicate with people with disparate knowledge in their native jargons, and mediate when they ignore each other's essential needs. While choosing, give priority to specialties where the rate of obsolescence of knowledge is high and to those that are susceptible to VUCA conditions.

Alternatively, since the decision makers' table isn't endlessly expansible and many equally important and competing factors determine who can be there, ensure that those present have access to these high-risk specialties. You may have to come up with novel ways of doing so.

For example, consider digital technology, which has a high rate of knowledge obsolescence. Consider creating a "fellowship" on your leadership team, temporarily occupied by a succession of digital natives. Or do what Unilever does in some European countries: Assign executives above forty "reverse digital mentors" who are under thirty. Everyone over forty isn't a technophobe, and everyone under thirty isn't a technophile. Nevertheless, the policy is brilliant. If reverse mentors were assigned to specific senior executives, it is possible that some of the executives would resent being singled out for being out-of-touch with digital technologies. Assigning reverse mentors to everyone obviates this problem. Moreover, it routinely exposes all executives to the worldviews of their younger staff.

Second, **allocate leadership responsibilities across your network.** You can't be everywhere and see everything. Distributed, thought-driven work can foment unmanageable VUCA conditions. The information you need will be buried in code, trade secrets, models, and people's heads. Reasons for actions may not be documented, and those for *not* doing something rarely are. Organizational boundaries, or sharp linguistic and cultural differences, exacerbate these challenges. People use easily available information, instead of better information that requires even minimal effort to acquire, so you may not think of seeking out hard-to-access information. (If you don't believe that, think of how many web pages you visit when searching for information online. Which ones?)

Centralized control will at the very least make you a bottleneck and will possibly make you a complete roadblock. Distributing leadership responsibilities will enable the leadership team to collectively navigate many more in-between spaces.

Catherine Cramton and Pamela Hinds offer another reason for distributing leadership when cultural boundaries are crossed.[10] They uncovered a complex dance of "cultural adaptation" during their study of bilateral project teams in three countries. Policies set in one country got modified by the other, which triggered reactions from the first, which generated more changes by the second, and so on. This process continued until both sides could live with the modifications made—or until the effort failed.

The adaptation occurred because each side's proposal didn't take into account the lived experiences—including those outside of work—of the other. For example, while flat organizations are valuable, even essential, in the digital epoch, in many countries, hierarchical titles have social implications that seriously affect people's lives away from their work. Distributing leadership, the professors wrote, can surface and help resolve such issues.

Third, **ensure that your people understand your strategic intent.** Strategic intent refers to the aligning of direction and pace of motion, instead of defining a vision (a desired, distant, end state) or a strategy (the grand plan for getting there).[11] It says, "Get to higher ground as quickly as you can," instead of specifying, "Go take *that* mountain, and only that one, right now!" In VUCA conditions, clear strategic intent works better. It gives the distributed leadership team the freedom to respond to local conditions within the framework of the overall effort. It is tantamount to saying, "This is where, generally speaking, we should go. Take whichever hill you need to in order to make progress."

Taking this approach will require a change in your behavior if you, like many executives, tell your people, "Don't come to me with problems. Come to me with solutions." Please stop! *Now.* Those words tell people not to share less-than-fully-formed ideas or not to tell you about potential VUCA events that are beyond their ability to resolve. Their silence may become your avoidable crisis. Say instead, "Come to me with problems. We'll reason together. I may push back or suggest a different approach, or I may take the task on myself."

Finally, wherever possible, **redesign the work your people do** to prepare them for the breadth they will need in their own careers. Doing so may also help you do your own job better: The "in-between" spaces may narrow, and many more minds may help determine how to fill them. Before her departure in 2018, Susan Sobbott was the president of global commercial payments at American Express and was responsible for 40% of the company's revenues. Asked how she decided someone had "high potential," she said:

> At some point I want my talent to have an expansive view for me to think of them as high potential. As President or as CEO, I can't run a business

all by myself. And you can't help me with niche expertise only. That's too much pressure on me and on you. I need you to see how different things fit together and how the business functions and I need your creative problem solving to bridge the gaps between experts.

While chapter 5 addressed individual and demographic diversity, this chapter addressed functional diversity. These issues are important in the digital epoch in and of themselves. They are also of critical importance for creativity (to be discussed in chapter 8). So chapter 7 will address how to bring such disparate groups of people together.

7 Being Truly Collaborative

The 1962 Cuban missile crisis brought the world to the brink of nuclear war. A brilliant book, *The Essence of Decision*, demonstrated that its unfolding couldn't be understood by considering the US and Soviet governments as rational, monolithic institutions.[1] That's because key groups nested within each government (e.g., the State Department and the Foreign Ministry; the Central Intelligence Agency and the KGB; the Defense Department and the Defense Ministry) had their own goals, policies, and processes that could run counter to those of the others.

That powerful insight is still relevant but usually ignored: The field of strategy addresses collaboration at the organizational level, while the field of organizational behavior addresses collaboration among people and teams. Neither takes the other into account. Collaboration in the digital epoch must be seen as a phenomenon that occurs at two interlocking levels.

The first level of collaboration, between organizations, has become more consequential than ever before. It is often essential for merely getting core work done, not just for reducing marginal costs or for increasing marginal revenues. Without it, a product or service might not even exist. Examples discussed earlier include Nokia and its battery partner, Boeing and the thirteen companies on three continents that created the hardware of the Boeing 787, airline alliances that enable seamless transfer of passengers, and consulting companies that field teams that span offices worldwide.

In the digital epoch, it is rare for a company to control every element of intellectual property it needs. Once a company sources intellectual property from another, the fates of both get tied together for long periods. Regardless of Steve Jobs's brilliance, without Corning's "Gorilla Glass," the iPod Touch, iPhone, iPad, or Apple Watch—and one of the world's largest market caps—wouldn't have existed. Very few other companies can create comparable glass. Apple could have worked with another company, but then its fate would have been tied to that company's.

Moreover, the initial choice of intellectual property can foreclose subsequent options.[2] A simple example is MacOS versus Windows. Until Apple replaced IBM-produced chips with Intel-produced ones, Windows software couldn't even be installed on Macs. Since then, each operating system can emulate the other, but performance suffers. Optimal performance requires software native to the operating system. Similarly, changing the vendors of software used to run businesses (e.g., ERP, CRM, and PLM) usually requires changes in policies and standards, procedures and practices, and even cultures.[3]

The second collaboration level is of the teams, perhaps from different companies, that work together. Each must adjust its work in response to what others do.[4] When multiple teams are present, "intensive" interdependence plays out; it is far more challenging than "reciprocal" interdependence between two teams. Professional (varying disciplines and skills), demographic or identity (varying genders, races, nationalities, cultures, and ages), and individual (varying behaviors) diversity add more challenges.[5]

The Global Survey showed that across the world, businesses executives don't collaborate well at the team level (see figure 4.4). Problems they mishandle, and opportunities they lose, can impact organizational collaboration. For example, in the Boeing 787 project, persistent problems with work at one facility forced Boeing to buy it out from a partner company.[6] Conversely, bad blood between organizations inevitably affects the teams actually working together.

Therefore, leaders of organizations that wish to collaborate need clear lines of sight to team-level collaboration. Without this, the many moving pieces in VUCA environments will inevitably produce major problems.

Developing the Collaboration Instinct

Asked what he looked for in potential partners, former Nokia chief pro-curement officer Jean-François Baril (introduced in chapter 4), replied: "A diamond in the rough. The technology is good enough, not neces-sarily great, but their people have the will to continue to fully polish it." How can you find and work with "diamonds in the rough?" Three issues are key.

First, **key people must be trustworthy and "relatable."** Blockchain technology records in public digital ledgers all transactions relevant to the provenance of items of interest. Social credit systems—the ratings people give online—can also indicate someone's trustworthiness. So tech-savvy young professionals often proclaim, "Digital technologies will make trust irrelevant." Their beliefs may be right someday, but for now, are simplistic.

Blockchains assure "transactional trust" in arms-length transfers of products or services ("A sold <this perishable food> to B after transport-ing it at <this temperature>."). Social credit systems offer insights relevant to comparable situations ("Should I offer delayed invoicing on shipping to a vendor with a two-star rating on eBay?"). Neither can tell whether someone is likely to collaborate well with strangers at an unknown future date to creatively resolve an unknown crisis that might strike.

A key challenge in collaborating effectively is overcoming concerns that someone's betrayal of trust will cause harm.[7] "Risky trust," when extended to strangers, is based on known capabilities—"This person is an experienced lawyer"—and known processes—"This is a reliable early warning system."[8] Nevertheless, such trust can fade very quickly,[9] and distance can magnify concerns.[10, 11] So experienced executives often assess the trustworthiness of their counterparts in other organizations and use that as a surrogate for organizational trustworthiness. Baril gave an example:

> Trustworthiness is about [a person's] core values. I was working with an Ameri-can company making a passive component. I met their head of sales for din-ner. He told me that he was the president of his church. I said I worked closely with people of all faiths and with atheists. He then added, "By the way, I don't

believe in God. I do it because believing in God is good for doing business in America." That I don't understand, I can't. That's where I have a tremendous issue. I told him, "How can I trust you when there is such a big disconnect between who you are in private and the face that you show in public?"

Inauthentic leaders can easily lose their credibility in the digital epoch (chapter 9 addresses this point). The American executive's in-person confession of inauthenticity eliminated the likelihood of real collaboration at a distance. He had sought to establish himself as worthy of trust by sharing a dark secret. Instead, he communicated that he would sacrifice any commitment if doing so served his interests.

As a variation on this, Susan Sobbott, the former president of global commercial payments at American Express, said that an executive's "credibility is directly correlated with relatability." On trips to corporate outposts, she would be "empathetic with team members and understanding of their situation." She'd talk about "their unique challenges about tactical or competitive points that were representative of their day-to-day experiences." She looked beyond "high intelligence—that's a baseline requirement"—and the ability to "synthesize lots of inputs and connect the dots" to whether they were

> good at collaborating with others in ways that consistently impressed me. Do they talk about connecting with colleagues, customers or direct reports? Do they proactively partner with others to execute and acknowledge their contribution? Do they have the resilience to overcome setbacks, resist feeling threatened, and keep going, to keep productive in the face of input or criticism?
>
> People who often complain about others or those who are "lone soldiers" wave "red flags" which suggest they can't trust and aren't trustable. Just like a growth mindset is needed to be open to new ideas, a collaborative mindset is almost always a prerequisite for high performance.

Compare this with the core attribute of scientific management—what does an individual do?—or even the core attribute of the quality movement—working collaboratively (primarily) with colocated colleagues. Not only must you collaborate, but you also must collaborate well at a distance—or you can't be a high performer.

Second, **recognize true collaboration is a strategic choice, not an inviolable ethical norm.** The prisoner's dilemma, a game-theory model, elegantly captures the concept of win-win collaboration.[12] Two people suspected of a crime are kept apart by the police. In exchange for a lighter—or no—sentence, if either blames the other, the police get the evidence to imprison the latter for a long period ("win-lose"). If neither breaks, they both get a light sentence or none at all ("win-win"). If both break, both get longer sentences than if they had kept quiet ("lose-lose"). A lose-lose outcome usually materializes because unless they talk with and/or trust each other, win-lose is the best option for each prisoner. If they are habitual (inept!) criminals who expect to partner indefinitely, the dynamics change. The likelihood of repeatedly facing the same choices makes keeping quiet—win-win—the optimal decision for both.

Win-win, the prisoner's dilemma suggests, happens under very specific conditions—communication, trust, and repeated interactions. Its pursuit can be valuable because working collaboratively, the parties to a negotiation may be able to identify benefits they hadn't foreseen initially. Expanding the set of possible outcomes with these unforeseen benefits could allow both sides to give up something the other party wanted while gaining something they valued.

Win-win has become a buzzword, however. People constantly announce their intentions to seek it even when there isn't any reason to do so. In the process, they ignore the critically important nuances mentioned above. Perhaps they believe that win-win is an ethical norm, wrongly conflating it with the common moral stricture to do unto others as you would have them do unto you. This creates another problem: Reciprocating a win-lose move with a win-lose move (i.e., a tit-for-tat) is more likely to ensure long-term cooperation than repeatedly responding to repeated win-lose moves with hopeful win-win ones. Stated differently, while "an eye for an eye" *could* "make the world blind," it could also prod people to collaborate.

Casual overuse of the term win-win makes true win-wins hard to achieve—for example, by precluding the tit-for-tat strategy. More importantly, overuse sharply drains the term's meaning. The concept of ethical fading (discussed in detail in chapter 9) essentially says that virtues professed abstractly from a distance become hard to maintain in the face of precise knowledge of real stakes. Baril metaphorically described this reality: "Most people collaborate like the kings of old, to show they are magnanimous. But the very first time their interests are threatened, they revert to being kings, doing what is best for them."

Worse, because of ethical fading, the human mind reinterprets actual behavior and wrongly remembers it as being consistent with the espoused virtue. Over time, as win/no-lose outcomes—personal victories that don't excessively annoy others—get called win-win, true win-wins become hard to distinguish.

If you truly value collaboration, choose your words carefully. *Words matter.* Don't speak of win-win when it isn't meaningful. Don't seek it when it isn't necessary. It is possible to reach equitable outcomes that don't bring with them unforeseen benefits. There is no ethical loss here. By limiting win-win's use to situations, people, and organizations where collaboration is important, you will be more likely to do the hard work needed to achieve it.

Finally, collaborative leaders **measure successes against their own prior achievements**. In collaborative endeavors, focus on creating a bigger prize than you could alone. If you do better than your expectation or your past performance, declare victory. Focusing on the division of spoils inevitably foments negativity.

Richard Parsons became chairman of AOL Time Warner in 2001, at a time when AOL and Time Warner executives were locked in an internecine battle that was destroying the company. He unified the factions and turned the company around when no one thought he could do so. Later, as chairman of Citigroup, he navigated its recovery after the 2009 global fiscal crisis. A consummate collaborator, he once told an interviewer:

> When you negotiate, leave a little something on the table.... I think people get hung up with their advisors, investment bankers, lawyers, and others, and every instance becomes a tug of war to see who can outduel the other to get the slightest little advantage on a transaction. But people don't keep in mind that the advisors are going to move on to the next deal, while you and I are going to have to see each other again.[13]

In the second decade of the twenty-first century, however, there is one *ethical exception* to this recommendation: It is appropriate only when the division of rewards is basically fair so that concerns about shares of spoils are at the margin. This is another reason why win-win should be seen as a strategic choice, not an ethical imperative. *When basic fairness is missing, win-lose—taking more at the expense of another—may be the truly ethical choice.*

Here's why: Societies worldwide have long deprived women and minorities of their fair shares of the benefits of success. Women often get paid less than men doing the same jobs (Organisation for Economic Co-operation and Development average 13.8% in 2016).[14] This is also true when they negotiate large contracts such as those of Hollywood actors.[15] In these situations, ignoring the sharing of benefits and focusing solely on prior achievements perpetuates injustice. Raising women's pay to achieve parity may keep some men from getting deserved raises, but that win-lose outcome isn't necessarily wrong.

Driving Collaboration at the Organizational and Team Levels

A baseline requirement for both the organizational and team levels is abandoning command-and-control leadership, the antithesis of collaboration.[16] Command-and-control leadership fails complex organizations hoping to innovate while collaboration serves multicultural, multi-organizational pursuit of innovation and creativity.[17] Assuming you agree and have a collaborative mindset, what could you do?

At the organizational level, **assess which network members should be partners.** Typical digital world networks include many organizations. Not all of them contribute equally: A company that sells a commodity is not the equal of one that contributes valuable intellectual property. Everyone doesn't merit equal treatment: The intellectual property provider may be a valued partner, while the commodity provider is easily replaceable. The intellectual property producer's contract may assure win-win compensation, while the commodity producer's may simply offer market price.

Deciding who does merit a win-win partnership isn't hard. Honestly answer a two-part question: Will we be hurt if they leave? Will they be hurt if we leave? Two "Yes" answers makes win-win essential, if and only if the other party also feels similarly. A "Yes" and "No," respectively, may lead you into a win-lose situation. A "No" and "Yes," respectively, isn't a license to be nasty; it should be seen as a reason for achieving a fair outcome (e.g., market price payment).

Moreover, when you talk to another company, do service-level and legal agreements dominate the discussions? Even in the best of relationships, you must occasionally consult such documents. But if you're tempted to always keep them handy, you lack the mutual trust partnership requires. Move on when you can.

At the organizational level, **elevate partnership norms and specify consequences of failure**. An organization doing what it contractually must isn't a partner any more than a stranger doing the bare minimum society expects. Put in place much higher standards whose breaches can't be easily explained away.[18] Be clear on the implications of their violation. Partnership must have not only privileges but responsibilities.

Baril gave an example that, stripped of idiosyncratic details, can be good model. His supplier network had an inviolable policy: Every member had a right to make a decent profit. Partners widely considered the measures of "decent profit" fair, even generous. When a new business head at a long-standing partner complained about heavy losses, his words challenged this norm. Baril:

> "This is terrible!" I said. "I make sure that all companies working with Nokia make more money with us than with anyone else. That is why I can ask for, and get, favorable terms, priority access to new technologies, etc. If you are truly losing money, I am obliged to end doing business with you. I do not want you to lose face in your organization." He asked for time to look into the numbers again. He returned to admit he was wrong and indeed, Nokia was his most profitable customer.

At the team level, **build trust and a shared identity early**. Trust takes three distinct, but mutually reinforcing, forms.[19] Purposive trust exists when people on all sides believe that their intentions are aligned and symmetrical. Cognitive trust requires needed capabilities to exist or be acquired. Procedural trust exists when requisite processes and systems needed to perform are or will be in place.

You must not only ensure that all three exist but also repeatedly highlight the fact that they do. Such positive championing is always useful. It is particularly critical for one variety of purposive trust—the inclusionary mindset, behaviors and actions discussed in chapter 5.[20]

Equally importantly, you must also talk up the importance of trust to the group's mission[21] and set and monitor group norms. Doing so creates a shared identity and carries the early trust forward.[22] Similarly, you must ensure early that people understand, or can readily access, answers to "who know what" questions about resources and expertise.[23] Lack of knowledge of existing resources can erode trust.

At the team level, **ensure people understand they are not their jobs**. Research on why smart executives fail spectacularly suggests that one behavior in particular hampers collaboration:

> They identify so completely with the company that there is no clear boundary between their personal interests and their corporation's interests.... Instead of treating companies as enterprises that they needed to nurture, failed leaders treated them as extensions of themselves. And with that, a "private empire" mentality took hold.[24]

Initially, such executives may believe disagreements with their organizations are attacks on them. Over time, this logic can reverse direction: They may characterize whatever benefits them personally as beneficial for the organization and whatever harms them personally as attacks on the organization.

Baril argues that a failure to separate people's jobs from their personal selves often turns work-related, issues-driven conflicts into relationship conflicts:

> Good leaders manage conflict well; they don't take it personally. You are in a *role*, say as head of marketing. People are not fighting against you, they are challenging the role, the interests, you represent.
>
> So, you need to respect, and earn respect as a person. You have to love people. I am not shy to say that. I ask a lot of questions about people. I do so because I want to know more about them. And if I don't respect them, I ask myself why.
>
> To lead a collaborative organization, you need to be able to defuse conflicts. You are not here to fight; you are here to find a solution.
>
> You are in a chair to do a specific job. You need to be able to stand up and look at the chair. Then there is no emotion. If you can't separate the chair from yourself, you will see everything as a criticism.

Amy Edmondson and and consultant Diana Smith have argued that mental processes similar to System 1 and System 2 govern how conflicts unfold in organizations.[25] The "Hot System" is impulsive like System

1, while the "Cold System" is thoughtful like System 2. Once the Hot System takes over, cycles of finger-pointing dominate. Breaking or forestalling this cycle requires the Cold System to reframe issues in ways that drain their emotional content. "Stand up and look at the chair" metaphorically captures this powerful idea.

At the team level, **create psychologically safe environments**. We've all been in situations where people felt powerless to speak up, or feared ridicule or punishment, or worried about looking ignorant, incompetent, intrusive, or negative. These situations weren't "psychologically safe." In them, people usually react by keeping quiet and trying to remain unnoticed. They hesitate to offer information, even key information they alone possess.[26] They feel particularly threatened when they feel isolated (a common problem when they aren't colocated) or when they hold a minority view.

Edmondson urges leaders to build psychologically safe environments. In addition to seeking out people's observations or ideas, they must "Embrace messengers…who come bearing bad news, questions, concerns or mistakes."[27] They should also give peers and subordinates a model to emulate by being "open about what *you* don't know, mistakes *you've* made, and what *you* can't get done alone" (emphases added).

Providing psychological safety isn't tantamount to overlooking unacceptable or repeated serious errors.[28] Edmondson urges leaders to consider the context. An avoidable failure that endangers life is inexcusable, and the consequences should be steep. On the other hand, small errors that cascade to create crises shouldn't necessarily cause heads to roll; they should trigger improvement efforts. For that, people must feel they can speak without retribution. If not, the problems will fester, and innovation and creativity suffer.

At both levels, **regularly assess and improve the health of your network**. Flat information flows benefit collaboration. Those who need to communicate should be able to do so. You must encourage this and create conditions that enable it.

Diagnose your situation by creating *network diagrams* that capture who communicates with whom, how, when, and why. Analytical software to do so exists, but when people at a whiteboard do so manually,

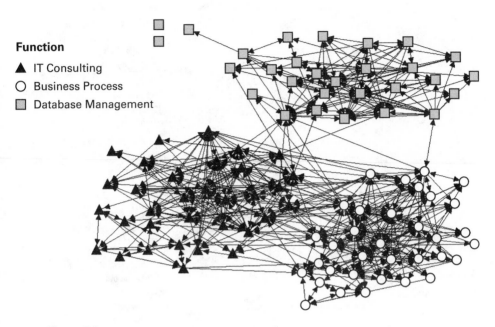

Function

▲ IT Consulting
○ Business Process
▣ Database Management

Figure 7.1
Network analysis of three groups in a company.
Source: Michael Johnson-Cramer, Salvatore Parise, and Robert Cross, "Managing Change through Networks and Values," *California Management Review* 49, no. 3 (Spring 2007).

they get motivated to apply the lessons. Different types of dots, colors, widths of lines, and uni- or bidirectional arrows can help represent different categories of people; conversation types; and when, how, and why the communications occurred.

Even simple network diagrams can help diagnose collaboration issues. Consider figure 7.1, which shows software-generated who-communicates-with-whom data from a real company. It shows three groups that communicate extensively among themselves: Most of the arrows are bidirectional. Two groups (IT Consulting and Business Process) have multiple people who communicate across group lines. Within all groups and across these two groups the interaction is reciprocal and even intensive.

The links to the third group (Database Management) are limited and primarily occur through one person. While the work may demand

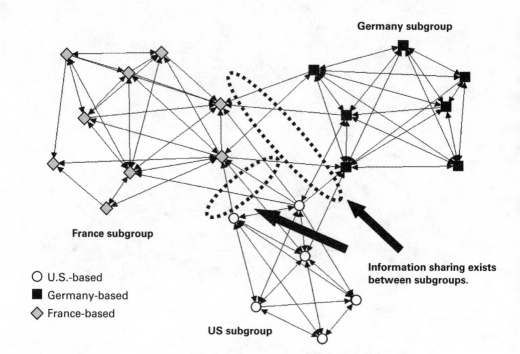

Figure 7.2

Network analysis of three geographies in a company.

Source: Michael Johnson-Cramer, Salvatore Parise, and Robert Cross, "Managing Change through Networks and Values," *California Management Review* 49, no. 3 (Spring 2007).

such centralization, this person could also be a command-and-control boss. Simple questions not captured by this diagram could determine the truth: What organizational benefit accrues from centralizing information handling? Could those benefits be acquired at a lower cost (or additional benefits be acquired at the same cost) by distributing the information flow?

Even without additional investigation, the diagram gives two indisputable insights that establish this network has problems. If the centralized point of contact in the Database Management group quit or got hit by a truck, the work of all three groups would suffer. Besides, the Database Management group has three isolated members.

You can also draw similar network diagrams across geographical regions (see figure 7.2). When work is distributed (Principle 3), if a geographical region is marginalized, visibility into VUCA conditions that originate there and access to unique knowledge housed there can be lost. A network diagram will give you insights into these potential problems.

Network diagrams can also help identify key people in the "informal organizations" that power most workplaces. They may have no obvious authority but may possess something far more valuable—credibility. These people energize and motive those around them. Their organizational knowledge "keeps the place running...in a way that is fair to everyone."[29] Among technical people, they have a "disproportionately large" impact on the success of knowledge work.[30] Do you know them? Are you protecting and rewarding them? You must if you're to be a truly effective collaborative leader.

Collaboration is critically important because the digital, VUCA world requires leaders to bring together knowledge, resources, and diverse peoples in the pursuit of goals. If you've done your job right, you'll be able to devolve authority for mission-critical endeavors even if you have to report—as Baril did—to the board. You'll be able to say, "We have a great distributed team in place and an amazing network of partners. They have all the processes and resources we need. They'll solve this problem. I don't have to look over their shoulders."

Equally importantly, collaboration is essential for the pursuit of creativity, the subject of the next chapter.

8 Championing Creativity

In 2017, I was coleader of an executive program for the senior executives of an Asian company with a storied past. The home-base country's former monarch had chartered the company over a century ago. That recognition allowed it to consistently attract great employees. Its operations stretched across East and South Asia, and revenues substantially exceeded a billion US dollars—a massive sum in purchasing power parity terms. Moreover, it was well integrated into the global economy: The leading European company in its industry and a global private equity firm had minority ownership positions. Nonnationals, indeed, non-Asians, held board and senior executive positions. So this story could be from most established companies worldwide, perhaps even yours.

The senior executives in the program visited a business incubator in the home-base country. The incubator's layout, available resources, and culture resembled those of incubators in cities renowned for digital innovation. Well-known entrepreneurs and venture capitalists had begun visiting to mentor or invest.

Serendipitously, one of the two entrepreneurs who hosted us knew four of the visiting executives. He welcomed them in culturally appropriate ways. The executives responded with affection and paternal delight. The young man had been a very promising manager in their company.

This entrepreneur's business model could transform how the visitors' industry sold products. In his presentation, he noted that he had developed its core ideas while working for them.

I asked both entrepreneurs a question with a predictable answer: Would they consider working for companies like their visitors'? I wanted the executives to hear their unhesitating response, which, as I had expected, was "Never." I also privately asked the ex-employee why he had left despite the mutual affection and respect. This being Asia, he chose his words carefully, essentially saying *he had been managed for productivity, not led for creativity. He had given the effort he was paid for, not the thinking he was capable of.*

As cerebral work displaces physical (Principle 4), "good work" takes the form of ideas and concepts. Executives must lead for creativity not merely as a nice-to-have virtue in R & D departments but (almost?) everywhere. Continued managing for productivity will put organizations—and their own reputations—at great risk. How, then, do you lead for creativity?

This chapter isn't about the tools, structures, and processes that Pixar or LG or Inditex or Hermès or Apple or Tencent or Fast Retailing or Google or other top innovative companies use to be creative. You can find those discussions elsewhere.

It also isn't about Design Thinking. If you don't know Design Thinking—the mindset, not just the methodology—you should rectify that deficiency immediately. Find a multiday workshop where you'll get hands-on experience. Good books can supplement, but not replace, experiential learning. Avoid lectures about it like the plague. I'll discuss some core Design Thinking beliefs since their absorption is key for the effective use of the tools they power.

Finally, it isn't about Steve Jobs's or Richard Branson's or Jack Ma's approach to creativity. You'll meet one genuine superstar but may not recognize him. His words will illustrate my recommendations. Those are based on interviews of, and personal discussions with, executives and professionals, as well as my own experiences of leading creative efforts.

But first, a handful of brief definitions: Though they are often used together, the terms "creativity," "invention," "innovation," and "Design Thinking" aren't interchangeable. Creativity is *the ability to look past received wisdom and traditional approaches to give form or structure to new ideas.* It requires framing—looking at—challenges in ways others haven't. It needs influxes of fresh ideas. It necessitates solving old problems in new ways, or new problems in old ways, or new problems in new ways—but never old problems in old ways. It demands testing solutions. If they don't work, it requires recycling to the start—or pursuing the unexpected outcomes. It's hard to define—and easy to recognize.

Invention introduces something new to the world. Innovations are (a) introductions of products or services that an organization didn't offer previously or (b) structures, processes, business models, or systems that didn't exist in their present forms in the present places in an organization. Most inventions don't become innovations, and conversely, most innovations, unlike inventions, adapt and adopt ideas from elsewhere.

Creativity is essential for both inventions and innovations. Design Thinking is a powerful tool for sparking creativity. Its core premise is that optimal design puts the needs of humans ("Is it usable?") on a par with the needs of technology ("Is it feasible?") and business ("Is it profitable?").

The Crushing Burden of the Past

It is easy to overlook the magnitude of the challenge of moving leadership of creativity to center stage. For over a century, businesses have focused on productivity (more output with less input) and process (one right way to act). Their leaders typically weren't personally creative and, most often, weren't required to inspire creativity in others. For eighty years, the boss was always right, even when he was wrong. That left little room for creativity (except in R & D labs if those existed in a given company). For another thirty years, non-R & D staff could be "creative with a small c"; businesses encouraged "continuous improvement" that made existing work error-free, faster, or cheaper but generally withheld the authority to do more.

A brief 2012 *Harvard Business Review* article urged limits on continuous improvement.[1] It argued that quality management tools created mindsets, metrics, and cultures that interfered with the "fundamentally different" needs of "discontinuous innovation." That this needed to be said a mere eight years ago suggests many executives and companies don't understand the needs of the digital epoch.

The challenges of learning to *lead* for creativity begin as early as business school. The most widely used means of educating leaders—case studies and lectures—are better for teaching analysis than synthesis or creativity. Universally taught concepts—such as discounted cash flows, optimization, and segmentation—ingrain a mindset of productivity. Mary Parker Follett published *Creative Experience* in 1924, but her outstanding scholarship (on a broad range of management issues) is largely forgotten. Charalampos Mainemelis, Ronit Kark, and Olga Epitropaki,[2] who did a 2015 review of the substantive body of research on leading creativity, established that its study—like that of teams—is a very recent phenomenon. Of the more than 300 articles and books they considered, 79% were published since 2001 and only 5.7% before 1990.

While top corporate leaders may like creativity's benefits, they aren't investing in it—yet. The Global Survey found business leaders don't effectively lead for creativity (see figure 4.6). A *2010* IBM survey reported that CEOs considered creativity their top requirement of leaders,[3] but IBM surveys since then either didn't replicate that finding or didn't report it. Other efforts haven't substantiated its results.[4] The IBM study would have been more robust if it had also asked whether the CEOs' priorities would change if finding leaders who led for creativity required missing profit or growth targets for a year or two. Even so, articles written almost a decade later routinely cite the lone IBM study. Finally, research suggests that creative people are *less* likely to become leaders.[5, 6]

These observations shouldn't be surprising. Experimental research suggests people espouse creativity but often reject creative ideas. In uncertain environments, they favor practicality over creativity. This preference adversely affects their ability to recognize creative ideas.[7]

Real leaders take needed steps even when these are costly; it's easy to be in favor of costless virtues. In 1987, Nobel laureate in economics Robert Solow famously opined, "You can see the computer age everywhere but in the productivity statistics."[8] Though he wrote at the very beginning of the digital age, his observation still holds.[9, 10] Yet, no economist would say, "Ignore computers." Similarly, today you'd be wrong to avoid leading for creativity. If you do, at some point in time, the blame for not doing so will, rightfully, fall on you.

What Isn't Enough but Can't Be Ignored

In the aforementioned review of the literature on leading for creativity, Mainemelis, Kark, and Epitropaki wrote that "creative leadership entails three alternate manifestations: facilitating employee creativity; directing the materialization of a leader's creative vision; and integrating heterogeneous creative contributions." Let's begin by summarizing these, adding context from fields beyond the study of leadership, highlighting important issues, and discussing where they fall short in the digital epoch.

"Facilitating" Leaders aren't creative themselves but enable others to be creative. They set appropriate goals, guide and direct idea generation, and evaluate and support idea execution.[11] They create needed work processes[12] and ensure effective communication.[13]

The study of innovation in perennially innovative companies offers many tools Facilitating Leaders can use for funding, team structures, staff evaluation, infrastructure, information flow, and key cultural practices.[14] Powerful, "multilingual" executives, who understand the work done by different disciplines, often lead major innovation efforts. They design team structures appropriate to the task at hand.[15] They try to recreate Pixar's brain trust[16] (a brainstorming session during which peers provide inputs and advice to a project team) or IDEO's Deep Dive (an embodiment of Design Thinking).[17, 18] They oversee handfuls of "clever people"[19]—valuable professionals who ignore organizational rules and have low thresholds for boredom—well.

Only a few of these tools (e.g., brain trust) were specifically designed for the digital world, but the insights of most are still valid or easily adapted. *Not* using them can make delivering creative outcomes difficult.

But as a complete toolbox, they are inadequate. Creativity is no longer limited to few areas. Leaders must engage many formerly "ordinary" people whom digital technologies have made far "cleverer" than they might have been in the predigital world. So the Facilitating Leader needs help.

Professions where creative work is inseparable from leadership—top executive chefs, renowned symphony orchestra conductors, and artistic directors of ballets—could provide insights. Such "Directing Leaders" provide the creative vision[20] and very precise instructions about execution to their professional staff. Those professionals—sous-chefs, line cooks, orchestra musicians, dancers, and others—must be highly capable merely to be selected to contribute. They must execute these directions perfectly[21, 22] since even minor failures are problematic.[23]

Early in the twenty-first century, several business schools in large cities began collaborating with local symphony orchestras to teach "Directing Leadership." One of the first to receive wide publicity used Ravel's masterpiece, *Bolero*, to illustrate the orchestra conductor's role. Executives heard maestros speak of clarity of vision, clear roles and responsibilities, coaching and feedback, and the importance of visibility to provide needed direction.[24] Some symphonies still offer such programs.

What they left out is as, if not more, important. Chapter 6 urged you to develop "a broad wingspan, not a long tail." Directing Leadership requires "a broad wingspan with *many* long tails." Directing Leaders have deep capabilities in multiple disciplines and lead highly skilled people with deeper capabilities in very few narrow disciplines. Consider, for example, the knowledge and skills of an orchestra conductor.[25]

While conductors need not know how to play every instrument in an orchestra, they must excel in at least one instrument, and ideally, more than one. They must, however, know the capabilities of each instrument and the specific challenges musicians playing them face. Furthermore, they must be experienced ensemble performers themselves so as to understand firsthand the ways musicians operate individually and in groups.

Long before approaching their orchestras for rehearsals (let alone performances), conductors spend countless hours mastering musical scores that can have 100 to over 600 pages. They have to reproduce in their heads how the musical notations from the score will sound; this activity is similar to the "voice" people "hear" in their heads while reading a beloved's email. The difference is conductors have to do so not just for each instrument individually but also for all of them collectively.

Conductors don't just present what is on the score sheets; they also interpret the music. To do so, they must master the intricacies not only of music theory but also of the context in which each composer was working, as well as how other conductors have interpreted each composers' music through the ages.

During rehearsal and performance, conductors must be able to distinguish among the various streams (e.g., brass or woodwinds), so they can catch errors or balance loudness or align beats. They must be able to get the entire orchestra to course correct on the fly if a featured soloist or opera singer makes an error. All of this must happen under stressful conditions.

Conducting, then, isn't just waving a baton. Much of this knowledge must be recreated for each symphony or opera in a conductor's repertoire. Indeed, over time, such work changes the shape of conductors' brains.[26] Expert conductors also develop better music-related long-term memories and divided-attention capabilities than even comparably expert pianists.[27] The knowledge and skills summarized here are essential for the clear vision, roles, responsibilities, coaching, feedback, and visibility conductors seem to provide effortlessly.

This breadth and depth of knowledge is an extraordinarily high bar for success modern businesses, which, unlike orchestras and restaurants, incorporate very diverse activities. How many separate bodies of knowledge must a Directing Leader master in the auto industry? Moreover, knowledge in many fields important to businesses—such as digital technologies and genetics—have high rates of obsolescence, while knowledge in the field of classical music has remained relatively static over centuries. That makes the challenge a lot tougher. A leadership model that's only accessible to extraordinary individuals isn't scalable.

Additionally, since the same piece of music involving the same musicians sounds different if the conductor is changed,[28] Directing Leadership only allows the leader's vision and creativity to flourish. This type of leadership *only* makes sense when the identities of an organization and its leader are inseparable; that wasn't true even at Steve Jobs's Apple. It certainly doesn't make sense when distributed work creates and combines unrelated intellectual properties.

Finally, Directing Leadership is inherently autocratic and can produce coercive workplaces reminiscent of the scientific management epoch.[29, 30] People working in them may experience psychologically unsafe conditions more often than those working in other workplaces. This reason alone makes Directing Leadership a terrible model, perhaps worse than thoughtful Facilitating Leadership.

Could Facilitating Leadership and Directing Leadership be combined into something better? Directors in the movie industry do just that. They serve as "Integrating Leaders" with key roles but must allow screenplay writers, actors, sound designers, and others to be creative in their own right.[31] If these nominally secondary roles don't collaborate and contribute, movies suffer. So Integrating Leaders must be firsts among equals, albeit by a big margin. They negotiate and collaborate, confront and compromise.[32] In their review, Mainemelis, Kark, and Epitropaki added:

> Integrative leaders have to be more facilitative than Directive leaders and more directive than Facilitation leaders. This does not imply...Integrative creative leadership is an additive function...Integrative leaders cannot abstain from proposing creative ideas [as Facilitative leaders do]...nor elicit [work that conforms rigidly to their vision]...as Directive...leaders do. Integrative creative leaders have little choice but to share with followers the "authorship" of a creative work in a way that neither Directive nor Facilitative leaders can do.[33]

Integrating Leadership eliminates the problem of lack of creative contributions by others. This ceding of power may even make the work environment less coercive: Because each member of the group makes a useful contribution, the power dynamics of the leadership team shift away from the Integrating Leader.

But the huge challenge of depth and breadth of knowledge remains. I learned this during a visit to one of the world's best special effects/computer-generated-imagery makers, the Weta Workshop in New Zealand. Among many other movies, Weta contributed to *The Lord of the Rings* trilogy. In addition to being commercial and critical successes, the trilogy collectively won 17 Oscars, including Best Director for Peter Jackson and four Oscars for Weta.

I asked a senior designer, "If Peter Jackson wants something and your expertise tells you that's not a good idea, what happens?" He described how detailed these conversations get. Bottom line: If top directors ask for something, like orchestra conductors, they know the field well enough to understand what effort, knowledge, and problem-solving will be required to deliver it.

In the creative arts, not far from directors are producers. Mainemelis, Kark, and Epitropaki's framework described them in passing, but very little of the existing research had focused on them. So the professors lumped them together with other Integrating Leaders. In reality, they are perfect prototypes for leaders of creativity in the digital epoch.

A Producing Leader of Creativity

Producers in the performing arts[34] define why something merits creation, shape the overall creative effort, hire and assemble all the talent (including the director), connect them to each other as needed, and ensure they collaborate effectively. Their ability to spark collaborative effort effectively allows the creativity of the talent to shine through—or not. They must understand the skills they need, but unlike the director, needn't be experts in everything. They must know enough to judge who is good in which field, and how to get the good people to work together.

In contrast to the practices during much of the twentieth century, in recent decades, the breadth of knowledge needed and the costs of production of movies have risen sharply. So the industry has begun practicing their own version of distributed leadership: Today's movies, unlike those of the past, have multiple named producers.

Effective "Producing Leaders" have the ability to navigate the in-between spaces. They have the Directing Leader's broad wingspan, but not necessarily the multiple long tails. They provide the Directing Leader's "creative vision," aligning the direction and pace of movement. Instead of engaging in command-and-control leadership, they assemble the right people and engender deep collaborations that reduce the potential for destructive conflict. They can effectively use the vast array of tools of the Facilitating Leader and may experiment and develop their own.

Ed Catmull is a great example. Catmull cofounded Pixar and retired as president of Pixar and Walt Disney Studios. He won an Oscar as one of the producers of Pixar's first hit, *Toy Story*.

Always interested in graphic arts and animation, Catmull turned to physics in college after realizing he lacked the talent to be a top artist. Realizing that a bachelor's degree in physics would give him only basic knowledge of that field, he earned a second major in computer science.

Laid off by Boeing, he sought a PhD in computer science at the University of Utah. That "unstructured" program offered "a safe environment for people to create" and "a safe place to make failures." For him, those lessons became "the right way to think." Other experiences in the program gave him "a new goal in life: to produce an animated film," a "dream" he achieved with *Toy Story*.[35]

Throughout his career he remained connected to academia, publishing his work and encouraging his employees to do so. Among other things, this helped him network with brilliant professionals. Early on, while running a research lab at a university, he realized a potential hire, Alvy Smith, was "more qualified for my job than I was." He "swallowed his pride and offered Smith" a key job at the lab. Smith became a close friend and cofounded Pixar.[36]

In the late 1970s, shortly after the release of the first *Star Wars* movie, George Lucas hired Catmull. Computer animation was in its infancy, and Lucas wanted to do more. Among others, he approached Catmull, whose computer animation of the human hand, done as a class project at the University of Utah in 1972, was one of the earliest accomplishments in the field. Catmull got the job because he honestly answered "an impossible question: 'Who else should we talk to?'" While other applicants "did the obvious thing" and "didn't give any names," he "gulped" and listed everyone whom Lucasfilm had already identified as candidates.[37]

So Producing Leaders

- Provide Facilitating Leadership using the vast array of available tools or by creating new ones.
- Provide "creative vision" like Directing Leaders in a form similar to the concept of strategic intent.
- Have a Directing Leader's broad wingspan, but not multiple long tails.
- Engender deep collaboration among the people they bring together.

  Additionally, as discussed below, they

- Have the ability to take the perspective of those not at the decision-making table.

- Are able to change their minds.
- Create conditions in which people ask for and give help.

A long tail is neither essential nor useless for Producing Leaders. Catmull's long tail in computer science had limited direct value at Pixar. Instead, his lifelong breadth of interests in the sciences and the arts and the vast network of brilliant people he had cultivated from multiple fields became key. He also kept connecting dots and learning about areas he didn't know about.

Table 8.1 captures how the Producing Leader differs from the three aforementioned models.

Developing the Ability to Lead Creativity in a Digital, VUCA World

Sooner or later, you'll organize a Pixar-created Facilitating Leadership tool, the brain trust. When you do, remember Catmull's description of his hiring Smith and his own hiring by George Lucas. They illustrate his deep commitment to two key brain trust principles, "There's no ego" and "Peers giving feedback to each other."[38] If you don't, instead of sparking creativity, this powerful tool will inevitably degenerate into a "stage-gate process."[39] Whereas the stage-gate process is a management review that makes go-no-go decisions, in the brain trust, "it's up to the director of the movie and his or her team to decide what to do with the advice."[40] The separating line is very thin and easily violated, as it once was in Pixar's own technical areas.

The prior paragraph should reinforce an idea from chapter 5: "You can't do the right thing consistently unless you actually care it is the right thing to do. *No set of rules would have prepared the coder to lead a team that included accomplished women.*" In other words, mindsets matter. That's why each chapter since that one has first addressed mindsets. What mindset do you need to be a leader of creativity?

First, **ensure you (can) empathize**. Years ago, my family was remodeling our house. We knew exactly what we wanted in our kitchen, but our very accomplished architect stuck with traditional norms. She reluctantly relented when we challenged her, "Do you cook? Well, we

Table 8.1

How the Producing Leader differs from the other models for leading creativity

Dimensions of Comparison		Facilitating	Directing	Integrating	Producing
Inputs	Leader's depth of knowledge	Nonexistent/Low	High	High	Medium-low (but learns fast whatever's needed)
	Leader's breadth of knowledge	Nonexistent/Low	High	High	High
Provision of guidance	Whose creative vision matters	Can vary	Leader's	Shared; leader is first among relative equals	Shared; leader is first among equals
	Locus of leadership	Centralized; loose links across power centers	Highly centralized	Moderately centralized	Distributed but aligned by strategic intent
	Leader's attitude	"The organization must get it done"	"I alone can get it done; you are replaceable"	"We can get it done"	"Without you, we can't get it done brilliantly"
How output is created	Leader's creativity	None/low	High	High	High
	Team's creativity	High	Low; subordinated to leader's needs	High	High
Conditions of creative work	Role of empathy	Unspecified	None/limited	Required	Essential
	Role of collaboration	Can vary	None	High/intensive, mostly two-way	Intensive multiway
	Ability to handle VUCA conditions	Low-medium	High-medium	High-medium	High-medium
	Leader as mediator	High	Low	High	Very high
Outcome and risk	Ownership of Outcome	Varies, low to high	Absolute	High/shared	High/shared
	Key risk	Process stifles creativity	Psychological safety	Weak collaboration, psychological safety	Weak collaboration
	Ownership of risk/consequences	Low	Absolute	High/shared	High/shared

do. So why don't you listen to us?" To her, *our lived knowledge* had no value. Weeks later, we rejected a top-of-the-line cooktop; moving heavy pots on and off the burners would damage its controls. We told the skeptical salesperson that engineers who didn't cook had designed it for purchase by people who used designer kitchens not to cook but to flaunt their wealth.

Our architect and those engineers didn't have to be great cooks. They did have to empathize with cooks: see the world from a cook's perspective and experience the joys of cooking and the challenge of handling large pots of boiling water. *Without empathy, they couldn't access the lived knowledge and couldn't create great works.*

Catmull's stories of the breakthroughs that created great Pixar movies are stories of empathy. Eight months before launch, the creative team for *Toy Story 2* hadn't figured out how to make a dilemma believable: Woody, the toy cowboy, was torn between staying back with the boy who loved him or going away with the toy cowgirl, Jessie. Catmull changed the creative team's leadership. The challenge

> ...was to get the audience to believe that Woody might make a different choice....[The new team] solved that problem by adding several elements to show the fears toys might have *that people could relate to*....the audience hears...in the emotional song...[that Jessie] had been the darling of a little girl, but the girl grew up and discarded her. The reality is kids do grow up, life does change, and sometimes you have to move on. Since the audience members know the truth of this, they can see that Woody has a real choice, and this is what grabs them.[41]

Design Thinking demands empathy, though it prosaically describes it as "field research" with experts. Practitioners must walk in their shoes before trying to develop solutions. If you know Design Thinking's core steps, but can't apply it reliably, perhaps your problem is empathy.

Genius, however, doesn't require empathy. If you are a singular genius, by all means, be a Directing Leader. But if you aren't, if you need others to contribute to creative endeavors, you must be empathetic.

The medical profession now understands that scientific and technical training can attenuate empathy. Studies show that more than two years of medical education produces a decline in empathy in medical students that remains till graduation.[42] Patients feel dissatisfied when doctors aren't empathetic, which adversely affects their health.[43]

Medical schools have adapted their curricula. They provide empathy training via storytelling and simulations. Ironically but naturally, digital technologies help. Virtual reality can recreate for a 26-year-old female fourth-year medical student the reality of a "74-year-old man with macular degeneration and hearing loss. A dark mass obscuring her central vision and the sound of muffled voices offer her a hint of what it's like to have these challenging health conditions."[44] Empathy scores are rising.

If a mere two-plus years of analytical, data-driven, scientific—and increasingly digitally intensive—medical training reduces empathy, no one is immune. As the kitchen redesign example shows, we stop empathizing when we "put on our professional faces." By mediating between us and the real world, digital technologies worsen the problem: They make us increasingly remote from others. We experience less directly.

Enhancing empathy is essential for bringing to the decision-making table the views of people who can't be there. If necessary, refresh your skills. As for other skills, learning requires explicit knowledge, modeling (observing and copying empathetic people), practicing, and getting feedback. If you are already empathetic, be a role model and provide feedback.

Second, **stop seeking uniformity and regularly challenge your own beliefs.** In the mid-1950s, psychologist Solomon Asch conducted experiments on conformity. He seeded experimental groups with aides who pressured test subjects. In one experiment, they wrongly insisted two lines were of different lengths; many test subjects quelled their doubts and agreed.[45] However, when asked individually, the test subjects gave the right answer. Even when subjected to pressure, the vast majority of them gave the right answer at least once. So the real lesson isn't that people conform, but under the right conditions, they can resist the pressure to conform.

Worshipping at the altar of process and productivity has bred the pursuit of uniformity into the DNAs of people and companies. This criticism doesn't mean medicines shouldn't be made in sterile conditions or cars manufactured to six sigma quality standards. It does mean that after addressing such legitimate needs, you shouldn't seek uniformity everywhere else.

You can help your people change. Instead of blindly following standard practices, ask them, "What could we do?" Add James Ryan's brilliant questions: "I wonder why...?" and "I wonder if...?"[46] Practice these on inconsequential issues, so you can apply them to consequential ones.

In addition, be open to changing what you think and do yourself. If you aren't, why should any creative person trust you? Ask yourself three simple introspective questions regularly: Which beliefs must I rethink? Which habits that made me successful must I unlearn? Which (new) capabilities must I (acquire)/relearn? Rethink. Unlearn. Relearn.

Practice resisting the temptation to tell. A leader who expresses definitive opinions and preferences too soon inevitably announces that everyone else's thinking isn't welcome. Doing so is particularly problematic when dealing with talented people. They will abandon you, and left with sycophants, you'll sink into groupthink. So learn to hold your tongue. Practice encouraging others to speak by asking simple questions: "What don't we know about this?" "How would someone in Company X (or another industry or country) have framed this issue?"

Creating Conditions for Creativity in Organizations

To lead people to be creative, you need to be inclusive to bring in a range of lived experiences (chapter 5) and diverse knowledge bases (chapter 6). Drawing on the lessons of strategic intent (chapter 6), you have to provide creative vision. You must foster collaboration (chapter 7). The next key issue is this: How can you create the conditions that foster creativity?

First, **redefine your own role to connect and encourage**. Andrew Hargaden and Beth Bechky's study[47] of six leading companies that do creative work uncovered four necessary conditions for groups to be creative: (a) People ask for help; (b) people give help; (c) they collectively reflect on, and reframe, challenges; and (d) the prior steps are reinforced, so they happen routinely and effortlessly. These conditions produce rich discussions that generate ideas that otherwise wouldn't have emerged. Ironically, many executives who desire creativity consider the first condition, help seeking, a weakness.

To help your people feel comfortable seeking help from, and giving help to, others, you need to start with yourself: Be a role model. Catmull addressed the importance of this:[48]

> [O]nce people get over the embarrassment of showing work still in progress, they become more creative....
>
> We make a concerted effort to make it safe to criticize by inviting everyone attending these showings to email notes to the creative leaders that detail what they liked and didn't like and explain why....
>
> The bigger issue for us has been getting young hires to have the confidence to speak up. To try to remedy this, I make it a practice to speak at the orientation sessions for new hires, where I talk about the mistakes we've made and the lessons we've learned. My intent is to persuade them that we haven't gotten it all figured out and that we want everyone to question why we're doing something that doesn't seem to make sense to them.

If you don't practice and encourage help seeking, why should anyone seek or offer help? Have you excised "Don't come to me with problems, come to me with solutions" from your vocabulary? If not, you're stopping people who need your help. Rethink everyday practices that worked brilliantly in the predigital epochs. An "open-door policy" is supposed to show you are a helpful boss but is useless for people based a half-world away. They work while you sleep. How will you address this reality?

Common questions in team meetings—for example, "Everything OK? When will you be done?"—send a powerful signal: "This is your responsibility!" Drop those questions in public. Do your network analysis and proactively point out who could help whom. You'll have succeeded when people ask for and give help online or in person without your pushing them to do so.

Reflection and reframing, a group activity, happens spontaneously and serendipitously. It thoughtfully replaces help-seeking questions with better ones that move the search for answers toward more fruitful directions. For example, at Design Continuum, a leading consultancy, one person's memory of a prior project involving intravenous fluid bags became the inspiration for the inflatable air bladder in Reebok's form-fitting Pump shoe.[49] At a major yogurt maker, the question "How can we increase sales 35%?" morphed into "How can we make eating yogurt fun?" The ensuing redesign of single-serve containers improved their visual appeal, made consumption easier, and enhanced sustainability; these produced the desired sales increase.

While saying "Let's collectively reflect and reframe" doesn't work, you can create the conditions that produce unplanned magic. Start with people with diverse expertise, lived experiences, and knowledge. Each may have experienced something or may know something or may remember something or may ask an outlandish question that turns out to be relevant to the issue at hand.

Make it easy for them to contribute so that "the private knowledge of individuals becomes the public knowledge of many."[50] To this end, enforce a lesson from improv comedy: No one's comment can begin with "but"; it must build on the immediately prior idea by beginning with "and." Add the other open-ended questions introduced earlier: "Could we...?" "I wonder why...?" "I wonder if...?" Forbid these eight dreaded words: "We tried that before and it didn't work." When goals change, conditions/context change, or available resources change, what may not have worked before may work beautifully.

In the digital epoch, distance is a key challenge. Those located far away may have very diverse experiences; you need to include them in reflection and reframing. A study of design and innovation in virtual teams spanned over four hundred people on seventy teams in multiple industries.[51] It found that unless a climate of psychological safety prevails in group communications, geographic dispersion, digital communication, staffing changes in dispersed teams, and national diversity (which includes identity and behavioral diversity) reduce innovation. So focus on what you have to do to improve psychological safety in such situations.

Second, consider whether you need to **change your sources of validation.** Not changing them regularly insidiously blocks creativity by stopping a broad search for answers before it has a chance to start. It has many faces; consider these two common ones: Do you often ask, "What are our competitors doing?"[52] Unless you look at world-class performance in specific areas, that question, the essence of benchmarking, conditions you to copy competitors. Do you routinely seek advice from consulting companies with strong industry practices? You may save money, because they know your business and "won't learn on your dime," but you squander more by losing your uniqueness.

Third, **create space and time for creativity.** Creative work needs focused time. Unless you are an orchestra conductor, multitasking or splitting attention can stymie creativity. Creative work may also require physical distance—going on a retreat—or psychological distance—abandoning the effort for a while to give time for ideas to percolate and gel in people's minds.[53] While resource and time constraints can drive creativity[54] and are even necessary,[55] unnecessary constraints can kill it.

Connect these ideas with the importance of help seeking and help giving. How can anyone give help spontaneously unless he or she has the space and time to do so? Worldwide, companies are stretching people beyond reason. A 2013 survey showed that American executives and professionals worked an astonishing 72 hours a week, mostly because of "useless meetings and emails, inadequate technology, disorganized or incompetent C-suites, and unclear decision-making authority."[56]

Take a hard look at the conditions you create for your people. We force people into cubicles or have them sit at shared tables. We constantly interrupt them via electronic media. We demand progress reports merely because we feel the need to be in control. We…you know this endless list already.

Fix what you can and explain why you couldn't fix the rest. Remember the lesson of procedural justice (chapter 5): Reasonable people accept decisions with which they disagree if they feel their inputs were considered and the decision-making process was fair. Also remember that this isn't a "once-and-done" exercise; you may be able to change in the future conditions you can't change now.

Fourth, **rethink how you hire people.** In general, hiring people who are your clones is never very good for creativity. Yet, executives routinely go back to their own schools to recruit the next generation of managers and professionals. They convince themselves that those schools must be amazing since they themselves attended them. Maybe they are, but different philosophies of education teach different approaches to tackling problems.

Jane Margolis's research on why American women stopped opting for computer science majors revealed another broadly applicable truth: *Aptitude and interest is more important than knowledge.*[57] When universities separated students who had prior coding experience from those who didn't, the latter got to acquire it. Many with aptitude quickly overcame their lack of experience and flourished.[58] A barrier identified for women also helped men who hadn't had prior opportunities to code.

Similarly, while "enough" knowledge is needed to fill open positions, executives often go overboard. Think of the last job posting you wrote. How many "must have done" and "must know" statements did you include? Why? Here's the standard against which you should measure your answer: Did the job requirements change from your specifications within a year? How quickly did your pedigreed new hire come up to speed at that time? If you truly had only an unavoidable handful, you pass. Otherwise, reflect on your motivations.

Guarding against overspecification is important. Those who may lack some knowledge but learn quickly are more likely to ask the "I wonder why…?" and "I wonder if…?" questions that spark creativity than those who uncritically accept received wisdom. As one of the most creative people in history, Einstein, wrote, "Imagination is more important than knowledge. For knowledge is limited, whereas imagination embraces the entire world, stimulating progress, giving birth to evolution."[59]

Finally, if necessary, **subvert the performance evaluation system.** The Global Survey asked the responding executives whether traditional performance measures were *less* relevant for thought-driven work. Across the world, 57% "Strongly Agreed/Agreed" they were. However, when asked whether the performance measures in their own companies captured thought-driven work, the same numbers (56%) said they did.

These two statistics disguised significant regional variations. In India and China, 62% and 48%, respectively, "Strongly Agreed/Agreed" that performance systems generally fell short. A further 9% and 27%, respectively, "Somewhat Agreed." In contrast, 80% and 75%, respectively, had no concerns about their own systems. Continental Europe continued its skepticism; 57% "Strongly Agreed/Agreed" and 18% "Somewhat Agreed" that performance systems generally fell short. Only 26% endorsed their own systems, while another 25% "Somewhat Agreed" they functioned effectively. The gap existed elsewhere, too, though it was smaller.

This exoneration of their own companies by all but the Europeans is hard to reconcile with the respondents' awareness of the problem. Cerebral, distributed work needs connection and inspiration,[60] and employee development, rapid change, and teamwork,[61] while traditional evaluation systems track and reward productivity. Indeed, starting a few years ago, many American companies have begun tackling this problem.[62] An extensive search (of English-language publications) revealed no similar reform efforts elsewhere. Clearly, this is a largely unrecognized problem.

Even among American companies that have made changes, replacement systems aren't necessarily better. They focus on separating compensation decisions from "weekly check-ins with managers to keep performance on course" and "quarterly or per-project 'performance snapshots.'" So they don't solve the core problem: A "performance review" won't produce "connection and inspiration"[63] regardless of whether it is done yearly, monthly, weekly, or daily, in conjunction with compensation discussions or otherwise. Check-ins to "keep performance on course" are still focused on productivity. They may be necessary but won't help you lead for creativity.[64]

Here is a simple, no-nonsense test you can use to assess the quality of your performance evaluation system: *Would it have stopped the Asian entrepreneur from resigning?*

Performance evaluation systems have a necessary, even legal, role. Unless you are the chief HR officer, you may not be able to change your company's. Moreover, the prior discussion of corporate leaders wanting creativity, but not necessarily investing in it yet, suggests you have to be courageous: Without doing anything unethical or illegal, subvert the performance evaluation system.

Regardless of standard practices in your organization, set your own people's goals by asking questions like these: Will this goal reward help-seeking and help-giving behavior? Will it promote collaboration in difficult situations? Will it help create a psychologically safe environment? Will it reward failures *that lead to understanding and insights*? Will it encourage exploration of new ways of working or identification of latent needs? In addition, switch goals across people who need to collaborate.[65] For example, give sales executives an operations goal and operations executives a sales goal. Since neither can succeed if the other fails, they will be motivated to consult each other. Instead of narrow, specific goals, set ones that will require rethinking how work is done and why. Who knows what such creative syntheses will produce?

As noted in chapter 5, Salesforce CEO Marc Benioff said, "[I]t's 2015 and the … business of business is improving the state of the world."[66] So executives will not only have to apply creativity to key business issues, but also to addressing the major challenges humanity faces. These efforts will be inextricably intertwined with the use of digital technologies. That simple, obvious fact will make them more challenging: As the next two chapters show, business has to sharply improve on its performance to date in the digital epoch. To do so, it needs "guide rails" for its creativity.

III   Guide Rails for Creative Efforts

The test of the machine is the satisfaction it gives you. There isn't any other test. If the machine produces tranquility it's right. If it disturbs you it's wrong until either the machine or your mind is changed.

—Robert M. Pirsig, *Zen and the Art of Motorcycle Maintenance*

9 Defining the Uncrossable Lines

In the Fall of 2018, I created a slide on technology-related corporate crises in the twenty-first century. I tried to minimize the problem, limiting my search to the largest economies. I included only four "How-can-you-possibly-ignore-them?" banks from the Great Recession. I selected only the largest data breach from a never-ending stream. I lumped together separate near-simultaneous crises at a company. I ignored companies that wouldn't be recognized by executives from their home countries.

Even so, I effortlessly listed twenty-five major companies globally. Some were nationally known and some regionally. Most were blue-chip global phenoms. At that point, I added the words "(and many, many more)" to my slide. My words didn't exaggerate: A 2017 survey of CEOs revealed that 15% had experienced crises in the prior three years and 30% expected "more than one crisis" in the next three years.[1] Early in the digital age, a meta-crisis of an unending spate of corporate crises is occurring worldwide.

Look across these crises, and it becomes clear that most were self-inflicted: A leader could have prevented each or at least kept it from ballooning from a problem to a crisis. Why are top companies struggling? Why is it that executives aren't anticipating possible negative outcomes of their decisions or are even actively creating them?

These executives, who spent/were spending/are spending their entire adult lives seeking the power, wealth, and prestige of top positions, cannot escape blame. All were guilty of deploying powerful technologies to pursue otherwise unachievable goals, while creating conditions that instigated or exacerbated problematic behaviors.

Beyond this, they fall into two groups. The first includes those who don't understand the enormous power of radical transparency (Principle 6). They are consistently unprepared for a high probability event: data misappropriation. Some even try to exploit Principle 6, unquestioningly tying their companies' fates to it.

Members of the second, much larger, group not only tend to regard social norms with disdain but they also abuse one or more of the other Principles. (Ironically, radical transparency usually brings their transgressions to light.) They distribute work without enabling the flow of key information. Or they rely on unseen, cerebral and discretionary effort without creating collaborative, learning-focused cultures. Or they bury decision-making in code without appropriate reviews. Or they don't distribute the authority to "pull the plug" when needed. Or they upskill work without providing needed training. Or they pursue emergent needs without essential safeguards.

Mark Zuckerberg and Sheryl Sandberg, the leaders of Facebook, are quintessential denizens of the first group. They ignored the power of radical transparency and didn't prevent data misappropriation.

Officially, the Facebook saga began in mid-March 2018, when the *New York Times*[2] and the *Guardian*[3] simultaneously reported on a data scandal with geopolitical implications. Facebook had given outsiders inappropriate access to data on 78 million Americans.

In reality, the saga began when Facebook's founders first sought venture capital funding. Early in the digital epoch, venture capitalists promoted a business model that enriched investors quickly: The median time to exit investments was a mere two years in 2001.[4] (It's more than four now.[5]) They asked entrepreneurs some variation of three questions: "How many eyeballs do you attract?" "What's the largest number of eyeballs you could attract?" "How quickly can you grow your numbers of eyeballs?" Selling eyeballs to advertisers brought in easy money. The mere anticipation of easy money unlocked corporate valuations that made early investors insanely rich.

From the moment entrepreneurs bought into that business model's logic, they had no choice but to pursue goals that would harm their users and the broader society. Whether they realized it or not, in effect, they had agreed to make their users private information public. Doing so couldn't but lead to a crisis. The only real unknown was "When?"

Facebook had to make screen time addictive,[6] so people would view more advertisements. It had to weaken privacy policies and make privacy modes hard to change, so users couldn't protect their own data.[7] It sold access to that data;[8, 9] doing so can significantly damage individuals[10] or make data breaches more likely. It made each of these decisions many times.

Facebook holds a patent that links an individual's access to credit to the social rating of the person's network.[11] Reflect on what this could mean to any marginalized group. It didn't use its technological brilliance to identify and remove patently offensive material about lynching, bullying, or "revenge porn." Doing more than the absolute minimum would have negated its legal fig leaf ("we are a platform, not media").[12] That stance enabled it to deny responsibility for posted content and assert immunity from governmental regulations and lawsuits for defamation or libel.[13, 14] But it also made it impossible to learn to use AI to counter massive misinformation campaigns. Even a *favorable* story on its belated efforts in this area coyly conceded that "compromises are made to accommodate Facebook's business model."[15]

Over time, leaders in Facebook's situation treat such crises as the cost of doing business. They aren't; they are the unavoidable outcomes of thoughtless leadership. Unsurprisingly, Mark Zuckerberg bristled[16] when Tim Cook said that Apple didn't "monetize" its customers because it believed privacy was a "fundamental human right."[17]

On March 30, 2019, as I was editing the penultimate draft of this chapter, Mark Zuckerberg published an Op-Ed in which he acknowledged the need for more regulation.[18] Whether this overture was serious[19] as he maintained or merely self-serving[20] as critics did will depend in large part on changes he is willing to make to Facebook's basic business model.

Facebook is merely a well-documented, broadly recognized example of the unappreciated real problem: All companies that share its business model also share its predilections. Unless they confront the dark side of radical transparency, in the near future, their stories may replace Facebook's in the headlines.

The second group of top executives, who overreached with digital technologies, includes Rupert Murdoch; the editors of his now defunct British tabloid, *News of the World,* authorized the hacking of a kidnapped teenager's smartphone.[21] For weeks, the police tempered its actions, believing an already murdered child was alive. In Ferdinand Piech and Martin Winterkorn's Volkswagen, executives and professionals collaborated to deploy diesel engines that cheated emissions-testing devices.[22] In John Stumpf and Carrie Tolstedt's Wells Fargo, branch managers oversaw the systematic opening of unauthorized accounts, the charging of unnecessary fees, and the falsifying of customer records.[23]

Regardless of the idiosyncratic details, each of these stories involved midtier executives who "pulled the trigger." They *were* blamed because they *could easily be* blamed. Subsequent investigations readily tracked their electronic fingerprints, sometimes detailing panicked fears of being caught. Their bosses, who created the reprehensible conditions that led to their actions, were the real culprits.

While an editor made the decision to hack, Murdoch and his hand-picked executives didn't set policies and processes for digital snooping in a scoop-focused media market—it went on for over a decade despite regular disclosures of problems.[24] While midtier executives and engineers misused Bosch's technology, Piech and Winterkorn's decades-long bullying (that is, scientific management epoch-style authoritarianism) gave rise to the plan.[25] While branch managers reviewed performance reports four times daily, Stumpf and Tolstedt set the goals, approved the performance measures, and encouraged or condoned the firing of those who complained.[26]

In each case, when caught, the top executives deflected blame. Murdoch and his son James presented an "I knew nothing" defense to a British parliamentary committee.[27] Stumpf argued he alone could fix Wells Fargo's problem; his replacement, Timothy Sloan, apologized for the bank's failures but not for the ones he made as as a top executive officer.[28] Twenty-nine months after his appointment, as withering political criticism continued unabated, Sloan finally resigned.[29] Though German prosecutors belatedly indicted Winterkorn, other senior Volkswagen executives who were Europe-based German citizens escaped the wrath of American regulators.[30]

The Columbia space shuttle blew up in 2003, killing seven astronauts. That crisis at NASA wasn't just self-inflicted, but also avoidable. For a little more than a week, shuttle program executives had ignored and even undercut the efforts of a handful of midtier experts who were trying to fend off the impending disaster.[31] General Duane Deal, a member of the US Congressional investigation board, famously said, "The foam did it...the institution allowed it."[32]

Today's spate of digital technology–enabled self-inflicted crises can be described similarly, but with an important addition: "The [digital technology] did it...the [company] allowed it...the top leaders created the conditions for it." The prime mover that drives these digital technology-powered institutional failures is leadership failure. To believe otherwise is to simply ignore overwhelming evidence.

Values in the Digital Epoch

The prior examples illustrated *the failure of human values to guide internal organizational choices. Values can also affect the functioning of organizational networks.* The next brief example, unlike the last few, is encouraging.

Jean-François Baril, the former chief procurement officer during Nokia's heyday (see chapter 7), had a simple mantra that went beyond Nokia's formal policies: Don't partner with organizations whose values trouble you. Forced to contract with a much larger brand-name organization that violated this mantra, he signed for the shortest possible duration. He told his staff to immediately start seeking viable alternatives. When the contract ended, Nokia was much bigger and could have negotiated better terms. Yet, Baril walked away. He said, "If I can't trust my counterpart to do the right thing when things are going well, how can I trust him to do the right thing when things get tough?" In a world of distributed, cerebral work, "things" inevitably and unexpectedly "get tough."

During that period, independent analysts routinely rated Nokia's network among the world's very best. Well after he left Nokia, Baril credited Pertti Korhonen, the former chief technology officer who had hired him, for the values-driven approach. Korhonen directed him to "do the right thing first, and things right after....What he meant was that you must have the nerve to say to the business…, 'I'm sorry, this isn't going to work this way.'"[33]

Values can also affect the effective functioning of networks of individuals. The support for inclusive leadership (see chapter 5) is itself an assertion of a specific value. Chapter 5 also urged leaders to "tackle the difficult task of assuring consistency across individual acts of inclusion." Shared values, of the sort IBM and J&J tried developing, can help assure consistency.

Finally, *values will determine how societies react to the truly widespread use of digital technologies.* In business schools, Chinese banks are often presented as exemplars for their embrace of digital technologies: In some cities, people equipped with accounts and appropriate smartphone apps from major financial institutions can effortlessly live without cash or credit cards. But that's only half the story. With full governmental backing, some banks are running *province-level* pilots of social credit systems (e.g., the rating you give your Lyft driver).[34] Those who fall foul of these can lose access to a range of government and private sector services, including car rental and rooms at good hotels. The banks are even using electronic billboards at rush hour to publicly shame individuals for their infractions.

Developing Values for the Digital Epoch

How can you shape values at these four levels—individual, organizational, network, and social? Talking up integrity or humility here would be pious, feel-good, and ultimately worthless. No business book can magically instill integrity in any adult who doesn't have it. Besides, with many more digital technologies yet to come, any such list of specific values would at least be presumptive, and at worst, totally wrong.

Research focused on the question "Why do some people who intend to do right still end up doing wrong?" suggests a more productive path. Those people start with good intentions but don't always act on them. After violating their espoused values, they convince themselves they were as true to the values as they could have been. So they talk the talk but don't walk the talk and then convince themselves that the walk was the talk. Understanding this problem, which researchers call "ethical fading,"[35, 36] can help leaders prepare, detect warning signs, and take corrective actions.

Before, sometimes well before, people make decisions, they espouse bright and sharply delineated values. They create broad, hazy scenarios of challenging conditions they could face. Focused on *what they should do*, they imagine making brave and idealistic choices. "My free social media platform will enable robust online communities." "My bank will serve all of every consumer's financial needs." "My car company will produce the best cars ever."

When they must make decisions, instead of abstract scenarios, people see the details of what is at stake. Their performance measures reward advertising revenues, profits, and rapid growth, not "communities," "serving," and "best." Realism and pragmatism justify actions that will get them *what they want*. Their espoused values become gray and easy to smudge. "My free service has transformed lives and our privacy policy discloses we own everything posted on our site. So, I'm entitled to sell their data." "We only want our employees to cross-sell because bundling credit cards with checking accounts will increase customers' purchasing power." "Passing that stupid Californian test will allow us to give Americans the powerful cars they love, not wimpy Priuses."

Later, people reflect on *what they should have done and actually did*. A range of mental processes convince them that they conformed to their espoused values. Others' actions, or situations beyond their control, produced any deviations. These myths become easier to sustain as precise details are forgotten with the passage of time. The selective memories and biased attributions and justification reset the baseline for the future. "Selling data funds a free service. Newspapers used to sell ads appropriate for their readers too." "We're helping build a credit history. One day they will thank us." "The regulators biased the competition to favor the Japanese. Anyway, everyone fudges these tests."

To recognize the onset of ethical fading and reimpose control, you have to practice identifying your "should self" and "want self." The following recommendations will help.

First, **keep your key values accessible; you can't apply them without recognizing their relevance.** In 2018, while waiting for a friend, two African American men at a Starbucks store asked for keys to the restroom.[37] The men were treating the store exactly as Starbucks wanted: as the "third place," after home and office.[38] Since they hadn't bought anything till then, the manager asked them to leave. When they didn't, she effectively treated them as vagrants and had them arrested. Radical transparency took over: Videos shot by other customers went viral, instantly turning a local outrage into a global one. Those who saw them forgot Starbucks's nearly fifty years of industry-leading corporate values. Many had done exactly what those men did without any adverse consequence.

Starbucks went beyond what other organizations have done in comparable crises.[39] In addition to taking responsibility, it changed its restroom policies. It shut down every US store for half a day to retrain every employee. Even so, it became the butt of comedians' jokes. Since then, most stories of discriminatory behavior by *other* organizations have rehashed Starbuck's ignominy.[40]

Three years earlier, Starbucks had launched an unusual initiative that encouraged store staff to write "#RaceTogether" on coffee cups.[41] Its leaders hoped to promote friendly discussions of this perennially difficult issue with their very diverse staff. Critics—digital technologies, naturally, played a key role—trolled them: No moralism with coffee. Chairman/founder Howard Schultz pulled the plug, claiming the termination was "as originally planned."

In a world where work is distributed (Principle 3) and/or cerebral (Principle 4), actions cannot always be seen, monitored, or controlled. When the Starbucks outlet in Philadelphia focused on an immediate "want" ("I need seats for paying customers"), it ignored a key corporate "should" and unleashed devastating consequences. #RaceTogether could have prevented this ethical fading by keeping society's racial problem front and center in the minds of employees. It might have even given them, the victims, and other customers the language to de-escalate the conflict and rectify the behavior.

After the deplorable event in 2018, I searched the works of journalists, op-ed writers, and comedians who were berating Starbucks for references to the #RaceTogether initiative. I couldn't find even one.

In the digital epoch, it is useful to remember that social media won't applaud the hard drudgery of keeping corporate values accessible and may mock you if you try. Even so, it will instantaneously disclose your failure to uphold those very values globally and destroy your reputation.

Second, **regularly seek constructive feedback for diminished self-awareness**. You can't always reliably self-assess your behavior in the digital epoch. VUCA environments facilitate ethical fading by obscuring the impact of bad behavior.[42] So do high stakes incentives.[43] People unthinkingly take more than their fair share.[44] This is easier when digital technologies obviate the need for in-person contact and/or provide anonymity. The spread of successful, but amoral, business models lowers the baseline for good behavior.

This is where a good mentor, even a peer mentor, can help by providing blunt feedback. ("You said you did <that> because of <this>; let me propose a different possibility...") Such conversations should be heavily biased toward improvement opportunities rather than praise. The best mentors are *objective...but biased in their mentees' favor*. Objectivity assures that they say to their mentees, "You really screwed up, didn't you?" Bias in their mentees' favor assures support for their mentees' efforts to redress and avoid problems.

Third, **create your ability to walk away.** A business school professor I respect wondered why "good" people didn't proactively leave companies that acted unethically. She realized that being responsible for their families' welfare often limited people's options. So she told generations of her undergraduate students to contribute to an "FU Fund" from every paycheck even before they followed the age-old advice of putting away money for retirement. Without an FU Fund, there might not even be an honorable retirement.

Being a Values-Driven Leader in the Digital Epoch

After protecting against ethical fading, what can you do on the job?

First, **decide where to draw the uncrossable lines.** Leaders are urged to be authentic. Authenticity offers many personal, social, and professional benefits while inauthenticity produces many negative outcomes.[45] This advice is easier given than followed. Leaders facing "must win battles" must outwardly portray confidence, even when they knows that failure is the most likely outcome. Are they being inauthentic? Whom would their groups rather follow—outwardly confident persons or those who acknowledge impending doom? To do their jobs, leaders often play roles, much like actors on stage.[46] What, then, is authenticity? Moreover, who decides when a leader is—or isn't—authentic, the leaders or their audiences?

The digital epoch magnifies this challenge. Leaders have to motivate those who may not have ever seen them in person. Besides, the balance of power shifts decidedly to the audiences. The leaders' inevitable digital bread-crumb trails, both contemporaneous and historical, can shape their audiences' beliefs and raise the specter of inauthenticity.

Authentic leaders specify *what, for them, is nonnegotiable*—lines they will not cross. These lines may differ from person to person, but they are bright, sharp, and backed by unassailable reasoning. They resist baits set for the brain's "want" mode.

Very few values should make this cut. If every value a leader holds is a line that can't be crossed, the collection of such lines would hogtie the organization. Conversely, if no value makes the cut, the leader effectively proclaims that everything is for sale.

Tim Cook has given a pitch-perfect demonstration of this recommendation. After publicly drawing Apple's uncrossable line, "Privacy is human right," Cook has taken on avoidable battles over the years in its defense.[47]

In 2016, the US Federal Bureau of Investigation asked Apple to unlock an iPhone implicated in a major act of terrorism.[48] Unlike many other technology companies, Cook refused, though the fear of terrorism could have turned public sentiment against Apple.

In 2019, several months after Cook challenged its business model, Facebook launched an iOS app that violated Apple's privacy policies: It could track users online without their knowledge.[49] Apple cut off all apps Facebook employees used for routine internal work. A few days later, Apple also meted out the same punishment to Google for similar privacy violations. Even though Cook relented several days later in both cases, Facebook and Google will probably try to reduce their dependence on iOS in ways that could impact Apple's financial performance. If that happens, analysts will point to these two actions and find fault with Cook's stewardship.

In the face of radical transparency, leaders who aren't authentic will be found out. At that point, they may still have authority but will have lost their credibility. In the digital epoch, you can't say a value is inviolable and not mean it. Cook's action did the opposite: Anyone who doubts his commitment to privacy can easily find these stories online.

Second, **make ethical fading easier to identify in distributed networks.** Having specified nonnegotiables, you should specify *what is critically important.* This list should be longer, but not by much. It will require protection against ethical fading.

How is "critically important" different from "nonnegotiable?" Apple's 2019 proxy statement[50] listed six corporate values: Accessibility, Education, Environment, Inclusion and Diversity, Privacy and Security, and Supplier Responsibility. Of these, it called two (Privacy and Education) fundamental human rights. By default the other four, while key, didn't rise to that level. There could be multiple practical reasons why. For example, all electronic products unavoidably use rare earths that damage the environment during mining, refining, and disposal.[51] They are also often mined in areas that have very troubling human rights records. Apple couldn't call protecting the environment a fundamental human right and still be in business.

Particularly in distributed networks, where in-person interactions aren't common, you should clearly specify critically important values in addition to nonnegotiable ones. An easy way to avoid ethical fading in connection with these is to regularly ask your people, "Did we do anything—or not do something—today/this week/this month that we would regret seeing on social media?" Without moralizing, this question reminds people of the power of radical transparency. It implicitly highlights the fact that *others may not necessarily share our version of reality.* Any hesitation should trigger a discussion; the incident may or may not be important. Any deliberate concealment, when discovered, should trigger disciplinary action.

You should also ask yourself this question: *What can we live without (for now)?* These are the trickiest. Because they are low priority, they are highly susceptible to ethical fading. Ignore them repeatedly and, over time, you will shift your baseline for acceptable behavior. So add this question: *Under what conditions will they become key?* If you can't answer this question, the espoused value merely exists for show. Again, consider Apple, so you can appreciate the distinctions. Apple under Tim Cook is a passionate, public supporter of LGBTQ rights, but it has to follow the law in countries that criminalize LGBTQ behavior. It can (and should) urge changes, but the very act of giving an LGBTQ employee benefits routinely given elsewhere could expose the employee to the authorities.

Third, **choose your words carefully when giving direction.** In chapter 2, I repeatedly wrote, *"Words matter,"* noting, *"They shape our thoughts, just as much as our thoughts shape them."* The words you use can have enormous, often unappreciated, power. For example, researchers have found that even when the payoffs are identical, naming a game "Wall Street" instead of "Community" leads to selfish behavior.[52]

When choosing words, pay as much attention to the context in which your words will be received as you do to the context in which you use them. Digital technologies magnify the importance of this recommendation, which predates the digital epoch. Once captured electronically, your ill-chosen words will go around the world in milliseconds and leave a bread-crumb trail forever.

Fourth, **resist the "bias toward action."** Slowing down action gives you time to evaluate how your "want self" may drive decisions.[53] You can rehearse how to respond to specific inducements that might produce ethical fading. Deferring implementation of decisions also helps; not focused on instant gratification, the mind shifts from the "want" mode to the "should" mode.

An executive interviewee does this well. Faced with complicated challenges, she describes them to her team but terminates the meeting without discussion. She reconvenes the team at least 24 hours later. The discussion is more thoughtful than it would have been had everyone been in a System 1 or "want" mode.

Finally, **frame issues to expose the "could," not just the "should" and the "want."** Should we do A or B? Decisions in ambiguous conditions can't be decided with data but must rely on values (see chapter 3). As they de-skill and upskill work or uncover new emergent needs, digital technologies force us to make many such decisions: In the event of an unavoidable accident, should a driverless car be programmed protect the driver of the car or to protect the lives of people outside the car?

Research suggests that sometimes "we should"—a values-based decision—may not produce the best outcome. Asking, "What *could* we do?" suppresses the either-or structure of the problem and frees the brain to explore options beyond those specified.[54] This simple change may even produce outcomes that satisfy multiple constraints.

Values may seem quaint, even archaic, in the digital epoch. Nothing could be farther from the truth. Spend a few hours with some good, near-future science fiction. You will come face-to-face with multiple realistic examples of how human values will shape the digital world for the better or the worse. What you do with that information is up to you. That said, there's very good reason why you should care. Chapter 8 noted that creativity "… necessitates solving old problems in new ways, or new problems in old ways, or new problems in new ways—but never old problems in old ways." Values matter because they help define the "new ways" you deem appropriate.

10 Developing Your Strategic Intent

As a leader in the twenty-first century, you will certainly have to lead for creativity to achieve organizational goals. You may also have to lead for creativity to help confront global challenges like the impact of AI on human workforces, climate change, gene editing, and the need to feed ten billion people by 2050. Digital technologies will play oversized roles in all.

Leading in either situation will require defining a coherent strategic intent. Recall that strategic intent aligns the direction and pace of motion across distributed teams. It allows distributed leaders to address local needs without centralized oversight while also ensuring that the local initiatives happen within the framework of the overall effort.

This is not happening with digital technologies. Within companies, R & D departments and business leaders are on completely different pages. For global challenges, researchers in academic and multinational institutions, CEOs trying to meet market expectations, entrepreneurs pursuing big dreams, politicians with many sincere and insincere goals, NGOs advocating for the voiceless and people at large must be corralled. These groups, each with its own interests, won't move in lock step. However, it would be helpful if they were not pulling in completely different directions.

An Extended Example: How Will AI Affect Humans?

Exploring the AI-workforce challenge can give a good understanding of what needs to be aligned. This discussion won't debate extreme outcomes—countless new jobs or mass unemployment. Instead, it offers the perspectives of many technologists and business executives, assembled from presentations at formal public events, articles, and personal conversations. Both groups have stakes in harnessing AI but also have very different points of view.

The Technologists' Perspective

Today's AI is very good at answering questions, but not asking them. One research trajectory focuses on understanding how people unconsciously do everyday tasks. Even so, generalized humanlike intelligence, popular from science fiction, is still many years away. Technology development is currently pursuing multiple domain-specific capabilities (e.g., subfields of health care).

Technologists are making progress faster than they expected. For example, the ancient Chinese game of Go is much more difficult than chess. In 1997, they predicted that it would take a hundred years to develop AI that could beat Go grandmasters. In 2014, that prediction fell to ten years. It actually happed in 2015.

So technologists feel constrained. They believe established companies aren't deploying AI systems quickly and broadly. Slow deployment is problematic because, inevitably, AI systems have access to more data after they have been deployed than during development. More deployments offer more opportunities to learn and improve.

Technologists expect young people and small companies to courageously ask brand-new questions and deploy AI faster. This will offer another benefit: Big technology companies won't be able to hoard AI-based power. Tiny start-ups using the right application programming interfaces to access the power of available AI systems (like IBM's Watson) will unleash unprecedented innovation.

Regulations will slow down efforts to improve the world. AI won't endanger human safety, and concerns about job losses are misplaced. It will augment, not replace, human capability and create "new-collar" (not white- or pink- or blue-collar) jobs that don't exist today. People will work the same number of hours, but on different things. They could also have much more free time, allowing them to lead more meaningful lives.

The Senior Business Executives' Perspective

Senior business executives believe AI's key value to companies lies in its ability to increase the performance of existing systems in marketing and operations. CEOs are being cautious and making deployment decisions incrementally; systems have to prove that they can deliver real value. Consequently, very few AI implementations have been at scale, and commonplace usage is far off.

AI will eliminate many jobs, but individuals and society shouldn't fear it. That's because, like prior major technologies, it will create even more jobs. A McKinsey Global Institute study[1] has forecast what could happen.

On the one hand, the study estimates that half of all current jobs can be automated and up to one-third of workers will have to change occupations. Its most likely scenario projects the displacement of 400 million workers between 2016 and 2030. People in advanced economies will bear the brunt of change.

On the other hand, AI and related technologies will create at least 390–590 million new jobs. In developed countries, these will come from increases in incomes and consumption and from greater spending on health care for aging populations. If countries increase investments in infrastructure and energy (and in a few other areas), the number of new jobs could be in the range of 555–890 million. Many jobs will be fields that don't exist today.

The economic transition will rival those that occurred as economies shifted out of agriculture and manufacturing. New jobs will require more education and midcareer retraining. Both businesses and society have to rethink how people should be educated and trained.

The Challenge of Strategic Intent

Both these perspectives implicitly assume that two realities of prior epochs apply in the digital epoch. In all past epochs, long-arc-of-impact technologies have destroyed prior jobs but created many, many more. Moreover, as the locus of the technologies was inside factories, the pace of business and societal change was gradual: Western economies, after all, embraced the quality movement epoch more than three decades after Japan did. In doing so, these perspectives gloss over four realities of the digital epoch.

First, while *past technologies de-skilled physical work, digital technologies also de-skill cerebral work* (Principles 1 and 4). Physical work was easy to transfer from farms to factories and across industries. The necessary retraining could be provided in days and weeks. Despite that, even advanced countries like the United States have struggled with people who entered the workforce without adequate primary education.[2, 3] In contrast, training for most cerebral work takes many years. Retraining for new sectors of the economy can't be done in days and weeks. The McKinsey report assumes retraining will be available in time to help the 75–350 million people, mostly in advanced economies, who will need assistance by 2030. Is anyone seriously preparing to provide this?

Moreover, half of all jobs being automatable is a lot. Its distribution won't be uniform. For many activities, it will produce unprecedented social turmoil. Pick any of the world's most crowded cities outside developed countries—Jakarta or Kolkata or Cairo. Every year countless people became professional drivers, embarking on a tenuous battle to reach the bottom-most rung of the local middle class. When driverless vehicles become commonly available, that rung will go out of reach for millions. Their lack of basic education will make retraining very difficult. Do CEOs and technologists have a responsibility to prepare society for this reality?

Second, *past technologies upskilled physical work, allowing many people to do with machines what few could without them. Digital technologies upskill cerebral work, allowing fewer people to do what many could* (Principles 2 and 4). Already happening in several industries,[4] this will accelerate as higher proportions of jobs become automatable. That will impose ever-greater pressure on midcareer job training.

Third, the digital transition is happening faster than any prior one. Jaikumar's research established that leading companies adapted to new epochs in about fifteen years and broad scale transitions across the economy took up to fifty years.[5] In contrast, McKinsey's research says the economies around the world must absorb unprecedented change in fifteen.[6] The problem—as the technologists' and senior business executives' perspectives suggest—is so far, the speed of technological change is swamping the incremental decision-making of CEOs. As such, the actual time to effect organizational and societal changes is being sharply compressed. The CEOs' incrementalism could be laudable if it were guided by thoughtful creativity; the focus on increasing the productivity of existing operations isn't comforting.

Fourth and finally, we haven't even begun to scratch the surface of the broader ethical issues digital technologies raise. For example, consider the unending demand for data. David Eggers's best-selling near-future science fiction book, *The Circle*,[7] describes the morphing of a good motive—radical transparency to expose political corruption—into a requirement for all people to constantly transmit their actions. *Could the need for data become a requirement to provide data?* It's not that much of a leap, really. It would require a line, a paragraph at most, in a user agreement for a bank account or car insurance or access to the Internet. China is already piloting social credit, a handful of insurance companies already monitor driving practices in real time,[8] and at least one company's "smart" televisions already track households' viewing habits without their *informed* consent.[9]

Because of these four realities, in contrast with prior epochs, this time we do need to ask, *What is the strategic intent?* How do we "roughly align" decisions being made independently by people who are oblivious to others' perspectives? How do we make sure that the missing perspectives (e.g., who is responsible for the retraining?) are taken into account? Far from seeking solutions, we aren't even asking the questions.

Business executives who wish to be called "leaders" can't duck this question. They can't simply respond to the future as it evolves—though that is a key skill—but must try to shape it. If not, they will be buffeted by uncontrollable events in a digital, VUCA world in which China's demand for copper affects the profitability of unrelated industries.

The first part of a framework that can help leaders defines Five Assumptions people unquestioningly make about digital technologies. Understanding these assumptions and making informed choices can open up "I wonder why…?" "I wonder if…?" and "Could I…?" paths for you to pursue. The second part of the framework addresses seven types of errors that confound major technological initiatives. Understanding these can help you set up guard rails for yours.

The Five Assumptions Made about Digital Technologies

The first of the Five Assumptions is the **Assumption of Benevolence**. Positive perspectives of digital technologies assume they will augment human capabilities; negative ones assume they will supplant or harm humans. The history of technology suggests both will happen unevenly.

Experts won't help you make good judgments. Knowledgeable people who are best equipped to be *skeptical optimists* about technologies default toward unvarnished, even gushing, praise. For its Fall 2016 issue, *Sloan Management Review* asked fifteen academics and expert practitioners, "Within the next five years, how will technology change the practice of management in ways we have not yet witnessed?" Ten articles exuded boundless optimism. An eleventh touched on a possible problem before turning optimistic. The twelfth, by an academic economist, neutrally discussed technical changes in corporate structure. The last two, also by academics, gently urged caution; one cowritten article raised concern about the ethics embedded in algorithms[10] and the other posited that "digital transformation needs a heart."[11] Twelve-to-two in favor of benevolence. No contest.

Average businesspeople are no different. In contrast to the conclusions of the aforementioned McKinsey report, they expect digital technologies to spare them the injuries they expect others to endure. In late 2017, online job search firm ZipRecruiter conducted a survey of one thousand job seekers.[12] Seventy-seven percent had heard the term "job automation," but only 30% actually understood it. Sixty percent considered the possibility of robots replacing humans at work overhyped. Fifty-nine percent of those currently employed *didn't* expect their jobs to be automated during their lives.

The Global Survey respondents answered two matched questions about increased thought content of work in general and in their jobs in particular (see figure 4.5). They largely agreed thought content was increasing, but their work was untouched. Clearly, they too optimistically believed they didn't have to change.

People unconsciously favor new digital technologies. They are more optimistic about *un*familiar than familiar digital technologies.[13] They blame technology failures on user errors, not the technologies.[14] Entrepreneurs continue investing in developing unsuccessful technologies hoping beyond reason for turnarounds.[15] In 2016, a publicly listed Finnish company, Tieto, put an AI system on its management team as a voting member.[16]

Awareness of—and mindfulness about—this bias can help you address it. Being skeptical about the proclaimed benefits of technology (without becoming a Luddite) is good. While early skepticism about new ideas kills creativity, judicious skepticism while evaluating those ideas produces better outcomes.[17]

The second is the **Assumption of Infallibility**. Algorithms already make critical decisions that affect people's lives. They require massive amounts of data for training. Here's a highly simplified explanation of the training process: An algorithm, fed data as input, processes the data and produces outputs. The outputs are compared to known correct results and the errors are fed back to the algorithm, which then adjusts its processing. After many such iterations, the algorithm makes sense of arbitrary inputs.

If bad or incomplete data are used during training, the algorithms learn the wrong lessons. This creates a chicken-and-egg problem: More data is available in actual use than during training, but deployment without full testing of the possible range of data that could be input during actual use can be dangerous. For example, if only Chinese features are used to teach the concept of a human face, the algorithm may not recognize people from beyond the Pacific Rim. If only Caucasian features are used, the algorithm may not recognize Asian or African ones.

Developers normally resolve this conundrum by relying on available data (e.g., Caucasian or Chinese faces, depending on where the development is taking place) instead of appropriate data (e.g., faces of all people from all parts of the world). Their decision usually isn't deliberately nefarious; people in other fields do the same too. Consider this very book: While arguing strongly for a global perspective, I've largely cited Western sources and/or articles published in English, relying on the interviews of executives with diverse backgrounds and the Global Survey to ameliorate this shortcoming.

With advanced digital technologies, the stakes skyrocket. With inaccurate data, development teams produce biased AI systems even when no individual developer is biased.[18] Explicit and implicit biases compound this problem. Current AI systems are poorly trained,[19] pose ethical conundrums,[20] and don't represent the population at large.[21] This problem won't abate soon. Once biased data get used, they corrupt development efforts everywhere. MIT Media Lab scientist Joy Buolamwini has talked about finding code in Asia that embodies racial biases normally found in America.[22]

Moreover, digital algorithms often produce great "intellectually" valid decisions that cannot be explained to nonexperts.[23] How do we regulate Facebook's algorithms? Or those of driverless cars?[24] Should we rely on algorithms for hiring? Police work? Stock trading? If yes, how would *you* explain to ordinary people that the world's economy depends on inscrutable algorithms? What would *you* say to justify AI systems that deny women jobs or recommend harsher sentences for people of particular races?[25]

Two researchers in ethics, Bidhan Parmar and Edward Freeman, described this challenge well:

> [T]he software code used to make judgments about us based on our preferences for shoes or how we get to work is written by human beings, who are making choices about what that data means and how it should shape our behavior. That code is not value neutral—it contains many judgments about who we are, who we should become, and how we should live.…
>
> Understanding how ethics affect the algorithms and how these algorithms affect our ethics is one of the biggest challenges of our times.[26]

Even when the algorithms aren't inherently inscrutable, there's another challenge. Since they aren't infallible, they should be subject to oversight. How could we balance oversight and protecting proprietary intellectual property? So far, corporate leaders have defied almost all oversight efforts. As the use of algorithms spreads and more grievous errors become public, the demand for regulations will rise. Where would *you* draw the line?

You need to take responsibility for the advanced technologies your company develops. Before you sign off on tests and deployments, you must assure yourself that they are safe, or at least their impacts are easily reversible. A good—definitely not perfect!—test would be *Would I authorize its application on someone I love?*

The third is the **Assumption of Controllability**. Even very knowledgeable people assume they can limit, or counterbalance, or control the evolution, or specific uses, of digital technologies. Reid Hoffman, executive chairman and cofounder of LinkedIn, wrote:

> [S]ome very smart people are worried about [AI's] potential dangers, whether they lie in creating economic displacement or in actual conflict...I am...backing the OpenAI project, to maximize the chances of developing "friendly" AI that will help, rather than harm, humanity. AI is already here to stay. Leveraging specialized AI to extend human intelligence in areas like management is one way we can continue to progress.[27]

In February 2019, OpenAI refused to release the full code for a program it had developed that could respond to prompts and write page-long, realistic essays, including creative writing. It explained its decision:

> Large, general language models could have significant societal impacts, and also have many near-term applications....We can also imagine the application of these models for malicious purposes, including the following (or other applications we can't yet anticipate)....Today, malicious actors—some of which are political in nature—have already begun to target the shared online commons...Due to concerns about large language models being used to generate deceptive, biased, or abusive language at scale, we are only releasing a much smaller version of GPT-2 along with sampling code. We are not releasing the dataset, training code, or GPT-2 model weights.[28]

Oh, the irony! An organization created to do good was concerned that one of its creations could be used for evil!

Christopher Manning, a Stanford-based AI researcher, disparaged the decision ("I roll my eyes at that, frankly"), noting, "Yes, it could be used to produce fake Yelp reviews, but it's not that expensive to pay people in third-world countries to produce fake Yelp reviews."[29] Manning trivialized the damage the program could do to people's lives and freedoms (just ask Mark Zuckerberg). Nevertheless, his criticism was valid: Someone is probably already creating a better program.

Three months earlier, a Chinese scientist announced he had used CRISPR, the powerful gene-editing tool, in an in vitro fertilization process that resulted in the birth of HIV-resistant twin girls.[30] He violated laws in many countries, but not China's. There, he violated a tacit agreement not to gene edit humans. While his effort might have been for good, it may yet trigger an arms race for fame and fortune that may not be benign.

But what then should we make of the story German pharma company Bayer posted on its website just thirteen days before the in vitro fertilization story? Hash-tagged *#CanWeLiveBetter*, it read in part

> November 2017 saw the latest milestone in our increasingly intimate and complex relationship with our genes, when 44-year-old Californian Brian Madeux was injected with copies of a corrective gene in an attempt to cure his Hunter's Disease. The injection was intended to directly edit Madeux's gene code, removing faulty pieces of the genome and stitch it back together.
>
> If successful, this treatment will be a major step forward for the medical application of gene-editing technology. But are we opening a Pandora's box? Now that we can, do we need to stop and ask whether or not we should?[31]

Underlying these three brief stories—OpenAI, the HIV-resistant twins, and the Hunter's disease cure—is an important question: *What sized "box" will contain our enormously powerful creations?* Too small a box and we forsake a lot of good. Too large a box and we invite potentially irreversible harm.

How will *you* make such decisions, which will often require collaboration with people outside your company? You'll sometimes be guided by law and *always* by your values. These decisions will be tricky; as chapter 9 noted, the logic of "everyone's doing it" can justify many sins.

The fourth is the **Assumption of Omniscience**. Renowned science fiction author Arthur C. Clarke formulated three "laws" of technology. One said, "Any sufficiently advanced technology is indistinguishable from magic."[32] His astute observation left a corollary unsaid: Throughout history, magic has been synonymous with power.

Digital technologies themselves, and their by-product—virtually instantaneous, seemingly infinite data on minutiae—engender faith that reaches levels associated with religion. They give an unwarranted sense of complete control. Just consider data analytics. Top universities,[33, 34] branded online education,[35] global NGOs,[36] and premier consulting companies[37, 38] all extol its ability to transform medicine, business, poverty alleviation, sports, entertainment, and countless other areas.

We can't forget these are just tools with their own limitations. How well will AI perform in VUCA conditions, which training data may not replicate well? We know that even in static conditions it fails to recognize people of some races.[39] Technologies will also fall short of what we expect. While AI can augment human productivity in creative activities, one expert equated this with "unleashing creativity."[40] The expert's opinion fell well short of the standard set for AI's ability to play chess or Go: "Win against a grandmaster."

Wide-scale use will expose more limitations. The Internet of Things will bring unimaginable benefits but will also heighten the problems we haven't solved with today's Internet—hacking, spam, viruses, and the like. Samsung's struggle to diagnose the cause of spontaneous fires in its Galaxy Note 7, a stand-alone device, revealed another critical challenge. Imagine millions of such flawed devices connected to the Internet of Things. Now add inscrutable algorithms—with no independent oversight—into the mix.

How do *you* feel about the power of digital technologies? When making decisions, how could you take into account their weaknesses?

The fifth is the **Assumption of Authenticity**. Authenticity has multiple meanings. It describes the match between a leader's real self and projected persona (see chapters 7 and 9). It also assesses whether someone or something is legitimate ("Is this authentic Thai cuisine?") or reflects the expected values ("Was the chef trained in Thai kitchens?").[41] These two latter meanings are relevant as people and robots begin working together.

Search for "people and robots working together" and you'll mostly find examples of mechanistic, utilitarian factory robots. In reality, robots may increasingly resemble sentient creatures, albeit with domain-specific intelligence. This will pose unimagined challenges for leaders.

Kate Darling argues humans are "hardwired" to anthropomorphize anything that seemingly moves of its own volition.[42] After playing with robot dinosaurs for one hour, test subjects refused orders to destroy them. Battle-hardened soldiers couldn't watch "wounded" robots continuing to do their assigned mine-clearing tasks.[43] People changed their behaviors in the presence of simpler robots that resembled self-mobile TV screens that projected images of noncolocated co-workers.[44]

In contrast, robots themselves have no emotion and (probably) won't anytime soon. Bots—software robots—that supervise people can "deactivate" (an official Uber term[45]) them unhesitatingly for whatever infraction their coding deems inappropriate. One journalist, after interviewing Uber spokespeople about their semiautomated process, wrote that if clear guidelines existed for infractions, they weren't known by the people affected by them.[46] Even so, Uber is rapidly introducing AI in all aspects of its business.[47]

In a legal filing, Amazon has described how its algorithm functions semiautomatically:[48]

> Amazon's system tracks the rates of each individual associate's productivity and automatically generates any warnings or terminations regarding quality or productivity without input from supervisors. [These] are required to be provided to associates within 14 days. If the feedback is not provided for any reason...the notice expires....While managers have no control over rates, they can override the notice if a policy was applied incorrectly....If an associate receives two final warnings or a total of six written warnings within a rolling 12-month period, the system automatically generates a termination notice.

Can a person complain against an unfair decision? To whom? Notably, Amazon's filing didn't discuss what happened to the "managers" who regularly overrode or ignored the algorithm's decisions.

Perhaps this lack of robotic emotion lies behind the research finding that people discounted music, paintings, and decisions supposedly created by algorithms.[49] They agreed these were comparable to those made by humans ("type authenticity") but questioned the algorithms' "moral authenticity" to produce them. When making judgments about ambiguous data, humans are also willing to be convinced by other humans, but not by robots.[50] In other words, *algorithms can do the work, but not be the work.*

The assumption of authenticity, beyond all others, will challenge you in the years ahead. It goes to the heart of what it means to lead. In humanity's entire past, people have led sentient beings—people and animals. *No one has led intelligent machines for whom humans can develop unreciprocated affection* (as in Darling's experiment).

You'll be wise to be skeptical—and empathetic. Consider a few obvious challenges you'll face. Should robots have legal rights?[51] Which robots should have which rights? If new digital technologies with uncertain or ambiguous characteristics are given human characteristics, people trust them more.[52] This knowledge will undoubtedly be used to peddle dubious products and services. Should it? If the mere knowledge of who did the work, person or algorithm, affects people's opinions, will people consider decisions made by machines outcome just? Procedurally just?

The Five Assumptions should inform the strategic intent of leaders. But first, let's consider another key issue: errors people make in complicated technological projects.

The Seven Critical Errors

Executives talk a good talk about errors but don't usually walk that talk. In the past, they treated all failure as anathema; today, many promote "failing fast," "failing forward," and "celebrating failure" seemingly without limits.

Real life is complex. Failures (and the errors that cause them) can be the source of profound knowledge when consequences aren't grave or they are controllable or easily reversible.[53] In other environments—as in operating theaters or at immigration checkpoints or in courtrooms or with key design decisions—like the Boeing 737 MAX's angle-of-attack sensor that caused two crashes—failure shouldn't be embraced in the name of innovation nor easily forgiven.

A granular understanding of errors can help leaders minimize the chance of catastrophic failure as they create bold human-machine systems. Recent research into this issue[54] suggests that people make seven types of errors: Believing something that isn't true (Type 1), not believing something that is true (Type 2), picking the wrong goals (Type 3), deciding before considering alternatives (Type 4), not acting when you should (Type 5), acting when you shouldn't (Type 6), and the combined effect of multiple small errors of the prior types (Type 7). Each type, discussed below, can affect digital technology initiatives.

Type 1 errors (believing something that isn't true) and Type 2 errors (not believing something that is true) are routinely taught in connection with data analysis in business, science, and engineering programs. Their discussion assumes we know the issues at stake (e.g., the possible errors in a training data set).

Reducing Type 1 errors unavoidably raises Type 2 errors, and vice versa. So, as a decision maker, you should err on the side of reducing the likelihood of the more damaging error. When you next have to choose, or sign off on investing in, a better training set for AI, ask, "Due to this investment, will the system give more credence to what is false or be less likely to learn what is true?" Follow up with "What are the key costs of its learning false facts? Not learning the truth?"

Type 3 errors (picking the wrong goals) occur long before then. In 2016, Microsoft withdrew its chat bot Tay for racist and other inflammatory output two days after giving it a Twitter handle.[55] Designed to mimic an American teenage girl, Microsoft wanted Tay to learn from interacting with real people. It chose…Twitter? A medium well-known for trolling? Which, until recently, refused to censor any speech? Even a small fraction of Twitter users could target Tay and corrupt it. That is *exactly* what happened. Adult supervision limits inappropriate behavior in children, even teens. Who was supposed to chaperone Tay on Twitter and how?

In the business world, Type 4 errors (deciding before considering alternatives) are well-known in theory, less so in practice. Good brainstorming guards against this problem by forbidding early criticisms of ideas. A British Design Council[56] document nicely illustrates (see figure 10.1) what should happen.

Divergent thinking (opening up options—illustrated as the flaring out of the squares near Discover and Develop) and convergent thinking (closing options—illustrated as the converging of the lines of the squares near Define and Deliver) should occur at two places in any project. Convergent thinking is important, but divergent is essential in

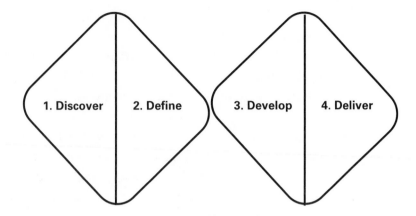

Figure 10.1
The double diamond phases.
Source: Design Council, "Design Methods for Developing Services."

VUCA conditions and for creativity. The former head of innovation of an Indian digital services company, an expert in Design Thinking, noted:

> Leaders who rush to conclude exclude the new possibilities.... Those that pause, suspend judgement and absorb diverse perspectives before zeroing in on the most feasible and effective ideas... succeed. Solutions... are more likely when they are built on the collective understanding of the emerging world, not based on a few individuals' extrapolated understanding of past beliefs and assumed futures.

Type 4 errors occur when divergent and convergent thinking don't happen appropriately. Microsoft's problem with Tay probably had its roots in an earlier successful launch of its Xiaoice bot in China. The project leaders simply replicated that effort, instead of returning to the drawing board.[57] Someone should have asked, "China censors police its internet. Twitter isn't policed. What could go wrong?" That should have led to divergent thinking.

The best leaders instinctively avoid Type 5—not acting when you should—and Type 6—acting when you shouldn't—errors. Today, executives working on technology initiatives are more likely to commit Type 6 errors than Type 5 because of speeded up competition and because they are blithely applying ideas and processes developed in simpler times.

A generation ago, when American businesses was under assault by Japanese businesses, *In Search of Excellence* taught American executives that the best among them had "a bias towards action."[58] In the 1990s, as the dot-com hype heated up, a method for field testing stand-alone software (like the early versions of Excel and Word) for bugs that would do no lasting harm became a product launch strategy with the sexy name of "minimum viable product." During 2000–2001, when software implementation initiatives—let alone development initiatives—routinely ran for multiple years before delivering meaningful results, a group of highly acclaimed software designers rightfully formulated "A manifesto for agile development."[59, 60]

The recent, avoidable Boeing 737 MAX disasters that cost hundreds of lives suggests how relying on these ideas *by default for all projects* could go wrong. A *Seattle Times* investigation revealed that with its development effort nine months behind that of the competing Airbus A320neo, Boeing's own internal safety analysis of its MCAS control system (that was supposed to prevent the 737 MAX's nose from being pointed too high and thereby stalling it)

> [u]nderstated the power of the new flight control system. … Failed to account for how the system could reset itself each time a pilot responded. … Assessed a failure of the system as one level below "catastrophic." But even that "hazardous" danger level should have precluded activation of the system based on input from a single sensor—and yet that's how it was designed.[61]

The article described numerous instances of short-circuiting of normal processes at Boeing and at the US Federal Aviation Authority in order to speed up the launch,[62] but that's only part of the story. The *Wall Street Journal* opined, "At the root of the miscalculations, though, were Boeing's overly optimistic assumptions about pilot behavior." It found that "Boeing assumed that pilots trained on existing safety procedures should be able to sift through the jumble of contradictory warnings and take the proper action 100% of the time within four seconds. … That is about the amount of time that it took you to read this sentence."[63]

The *Journal* also wrote, "The company reasoned that pilots had trained for years to deal with a problem known as a runaway stabilizer … [and the] correct response to an MCAS misfire was identical. Pilots didn't need to know why it was happening." Boeing wasn't alone in making this flawed assumption; the FAA had its own version of it. Currently, "FAA rules typically assume 'the human will intervene reliably every time.'" After the crashes, the FAA is rethinking its "reliance on average US pilot reaction times as a design benchmark for planes that are sold in parts of the world with different experience levels and training standards."[64]

During the design process, the MCAS was initially designed to move the tail fins only 0.6° out of a physical max of almost 5°. However, when test pilots worked with the advanced prototypes, this got increased to 2.5°. While "it's not uncommon to tweak the control software,"[65] this large increase was neither fed back to the designers, nor documented in the safety analysis submitted to the FAA (or in information provided to any airline).

In part, the flawed decisions were made because: "MCAS wasn't seen as an important part of the flight-control system.... [A]round 2013, the plane maker described the system as simply a few lines of software code."[66]

In part, the flawed decisions were also made because of other decisions taken far from the R&D labs:

> The assumptions [that pilots could react almost instantaneously] dovetailed with a vital company goal. To make the plane as inexpensive as possible for airlines to adopt.... At one point around 2013, Boeing officials fretted the FAA would require simulator training, the person involved with the plane's development said. But the officials, including chief MAX engineer... opted not to work with simulator makers to simultaneously develop a MAX version because they were confident the plane wouldn't differ much from earlier 737s.... It was a high-stakes gamble," this person said.... The company had promised its biggest customer for the MAX... it would pay it $1 million per plane ordered if pilots needed to do additional simulator training...[67]

In the original design, there was an alert feature which could have warned the pilots the MCAS was malfunctioning.

> Trouble was, that alert feature wasn't activated on MAX jets operated by Ethiopian and many other airlines. A contractor had made mistakes in software meant to activate them, but Boeing had told only certain airlines.... Boeing, which maintains the alerts aren't critical safety items, instead billed them as part of an optional package.[68]

This is far from the complete story of what happened and indeed, it may change in the months and years to come. It is also clearly an extreme case of a crisis. Even so, it offers a critically important lesson.

Executives make huge assumptions—on human behavior, what customers will do, how an engineering system will work, what the operating environment will be, and many more—when their companies create products and services with embedded digital technologies. They don't necessarily ask themselves whether their existing corporate processes and systems can respond adequately to the issues that could arise in a new epoch. In particular, there's a real danger in the fact that though most of today's digital projects are routinely delivered in days or weeks or months, executives continue to display their bias for action by pursuing agility with demonic intent while launching really complicated minimum viable products.

Calling out this serious problem is not an endorsement of "paralysis by analysis" or the glacially slow "waterfall method" (linear progression through requirements, design, implementation, testing, and maintenance) of software development. Nor does it suggest that rapid prototyping is detrimental—it is essential and indeed, its value is often underrated. It does imply that speed shouldn't be at the expense of thoughtfulness; even advanced prototypes shouldn't be released without careful consideration, if at all; and all necessary (not minimally required) testing should be completed.

Samsung's Galaxy Note 7, Boeing's 737 MAX, esoteric financial products, and many other similar major errors should convince us that, *in a highly connected, digital, VUCA environment, the cost of acting too soon (Type 6 errors) can be much higher than the cost of failure to act (Type 5 errors)*. As I've written elsewhere, "If the speed and cost for fixing errors and miscalculations are acceptable, by all means proceed with agility, aim to be first to market, or launch minimally viable products. Otherwise, steel your backbone and demand thoughtfulness."[69]

Finally, Type 7 errors result from a cascading of multiple small, individually inconsequential, errors of the prior types. These can combine to produce crisis-level outcomes. Don't downplay small errors! Instead, ask, "Could these cascade in VUCA conditions? Under what circumstances

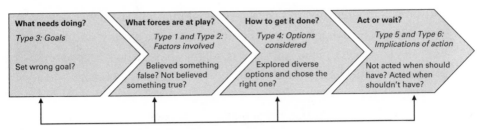

Multiple independent forces at play
Type 7: Cascading errors
Did small errors combine and magnify?

Figure 10.2
The seven types of errors.
Source: Lightly adapted from Mark Meckler and Kim Boal, "Decision Errors, Organizational Iatrogenesis and Error of the 7th Kind," *Academy of Management Perspectives,* published online October 15, 2018; in press.

and with what impact?" Moreover, set standards for risk before you begin. Tighten them if experience suggests, but don't weaken them. Over time, ignored risk standards inure decision makers to greatly enhanced risk.[70] Preprescribe steps that must be taken if these standards are ever breached.

Figure 10.2 illustrates the errors in the context of the evolution of a project. Note that the numbering isn't sequential. Type 1 and Type 2 errors are broadly used terms with roots in statistics; they were identified long before the others.

Formulating Your Strategic Intent

Digital technologies haven't yet lived up to the hopes attached to them. I don't agree with much that venture capitalist Peter Thiel says, but he captured this shortfall eloquently: "We wanted flying cars, but we got 140 characters."[71] More prosaically, while 2 KB of RAM in one spacecraft got humanity to the moon, 2 GB of RAM (a million times more) in each of billions of smartphones worldwide hasn't produced comparable wonders. Somehow we got sidetracked into friending strangers we wouldn't recognize if our lives depended on doing so.

The biggest challenge leaders face isn't technology but their own mindsets. The path of incremental efforts that many CEOs are currently pursuing is the easy one to take since it continues the last two epochs' focus on productivity. After all, their companies are already set up for this, and it appeals to financial markets. All they have to do is to focus on de-skilling (Principle 1) and upskilling (Principle 2) and automate existing operations.

In 2017, I visited a major global digital technology consultant's showcase lab for digital technologies. The consulting company did cutting-edge research there and also used it to convince senior client executives to spend tens of millions of dollars on digital transformation. The top featured use? Using virtual reality to speed up the training of equipment maintenance staff who are responsible for expensive capital assets.

The digital epoch demands bold, creative initiatives. For these, Principles 1 (de-skilling) and 2 (upskilling) may be useful, but Principles 3 (distributing work), 4 (cerebral work), and 5 (emergent needs) will often be key. Contrast the prior use of virtual reality technology with its use to give a young female doctor-in-training a visceral understanding of what it feels like to be an old male patient with a debilitating illness (see chapter 8). One can save some money, the other can help transform medicine—and make money. Which company—one speeding up maintenance training or one transforming medicine—would *you* want to lead?

Aligning your distributed leadership team's direction and pace of motion enables the flexible addressing of local issues within the context of a larger effort. The discussion of the technologists' and business executives' perspectives at the start of this chapter should have suggested its importance: Although digital technologies are transforming long-standing social contracts among people, between institutions and people, and between governments and institutions, the two key groups—and governments, regulators, and NGOs—are behaving like ships passing each other silently at night.

Consider formulating your strategic intent collaboratively with your leadership team and subject matter experts. Why? An adage long associated with Design Thinking is "Nobody is as clever as everybody." The digital, VUCA world is convoluted, and you may miss key issues which others don't.

Start with the mindset that everyone needs to adopt with digital technologies. *Bold, creative initiatives will require your replacing a productivity mindset with a creative one.* Mindsets are easy to ignore because the term is touchy-feely soft. ("OK, but what do I *do*?") You will *do*, but take a few moments to *think*: I wonder if...? I wonder how...? How/what could...?

The mindset change you need to make now is conceptually no different from the mindset changes your predecessors had to make during the last three epochal changes. They went from *precision focus* to *clearance focus*, from *machine focus* to *people focus*, and from *acceptable quality of individual workpieces* to *time-phased control of production variation*. These mindset changes were their biggest hurdles. Don't underestimate yours.

At the transitions of prior epochs, the new foci didn't eliminate the old ones, but supplemented them (and yes, pushed them to positions of lesser prominence). So the focus on clearance, for example, didn't mean micrometers were abandoned; it simply means micrometers were used where necessary and not by default. Similarly, in the digital epoch, your focus on creativity won't mean that you don't have to care about productivity; it does mean that wherever creativity is needed, it—not productivity—should be the default option.

Then consider your ideas for creativity, inventions, or innovations. View each through the lens of each the Five Assumptions. They will help you shape the response to your "I wonder if...? I wonder how...? How/what could...?" questions. What opportunities and issues arise? Where is flexibility needed? By whom? How much? Where must you be in sync? Why?

Finally, use the seven errors to identify where your risks are the greatest. Given the newness of many digital technologies and the intricacies of the VUCA world, pursuing bold goals will almost inevitably expose you to greater risks than automating existing tasks. Ask about them: "Is this risk worth taking?" For most, the answer will probably be yes. Then ask, "How can we protect against it?"

An executive officer once told my class, "Given the choice between doing something small and doing something big, pick the big; it will only take a little bit more effort." Good words to keep in mind.

IV Where Next?

Where the mind is without fear and the head is held high;
Where knowledge is free;
...
Where words come out from the depth of truth;
Where tireless striving stretches its arms toward perfection;
Where the clear stream of reason has not lost its way in the dreary desert sand of dead habit...

—Rabindranath Tagore, Nobel laureate for Literature, 1913,
Gitanjali

11 Building a Personal Leadership Philosophy

This book has covered a lot of ground. This final chapter is structured to help with the task of deciding which ideas you should assimilate now and how, or stated differently, "With the understanding I have gained, where do I go?"

Let's start by having you explicitly document what I call a "personal leadership philosophy." The personal leadership philosophy is an "owner's manual." While it's primarily meant to help you reflect on how you lead, you can use it to introduce yourself to compatriots.

The Prioritizing Exercise, Part I

1. In chapter 4, I alluded to the data Michael Watkins and I informally collected that showed that the average midtier executive spent two to three years in each role. Your experience may be somewhat different in terms of the number of years you have spent/are spending in each role. Regardless, which handful of principles of leadership do you adhere to across all your roles? In your current context, which additional ideas must supplement these? What will people remember about you when you finish your two to three years in your present role and move on?

   Take some time to create your personal leadership philosophy by answering these three questions. They aren't about your achievements but about how you lead on a daily basis. Begin each element with "I believe..." Add "Why?" Keep it short, one page at most in total. For now, ignore ideas from this book except one: If appropriate, sort your selections into the nonnegotiable, critically important, and can-live-without-for-now buckets.

2. Consider the corporate failures described in this book or others more relevant to your company/industry. Had you been involved, would your personal leadership philosophy have helped you function creditably? Could anything have stopped you from adhering to it? What lessons can you draw? Does your personal leadership philosophy need to change?

The Essence of the Producing Leader of Creativity

Let's consider a relatively complete picture of a Producing Leader of Creativity by summarizing the mindset, skills, and actions spread across all the prior chapters.

Producing Leaders of Creativity understand that their environment is volatile, uncertain, complex, and ambiguous and that digital technologies interact with these conditions in very specific ways. They are comfortable with, even embrace, a world in which China's demand for copper affects the profits in an unrelated business.

They embrace a creativity mindset and discard twentieth-century policies and mindsets that are no longer appropriate for a hyperconnected world. Recognizing that the baby steps many CEOs are taking aren't enough, they are bold in their pursuit of new futures. That said, they understand and act on the Five Assumptions and recognize that digital technologies can have unforeseen consequences. Their strong values and vigilance against ethical fading helps their organizations avoid crises that are swamping businesses globally.

They value breadth of knowledge, imagination, and fast learning. They take on the responsibility most others avoid—of filling the chasms that lie between narrow bodies of expertise. They question what they know personally and are willing to unlearn, rethink, and relearn.

They understand that they can no longer be dispassionate facilitators of creative efforts undertaken by a handful people in small playpens. As such, they become contributing members of the creative community. Moreover, instead of staffing their teams solely with people selected on the basis of what they already know, they seek out those who are always willing to learn.

They are inclusive and empathetic. They configure their performance systems to attract diverse creative people. They build cross-cultural trust and make efforts to connect people.

They aren't autocrats who hear only the voices inside their own heads; they seek inputs from others. To this end, they structure partnerships that people want to join, distribute leadership responsibilities, and align efforts by defining the strategic intent.

They ensure information flows freely through their network and institutionalize people's ability to ask for and give help. Having mastered the art of asking "I wonder if…?" "I wonder how…?" and "Could I…?," they provide psychologically safe environments in which their people can answer these questions honestly and pursue creative outcomes.

The Prioritizing Exercise, Part II

1. Evaluate yourself against the profile of a Producing Leader of Creativity. Where do you do well? Where do you fall short? Feel free to add from prior chapters elements that might be missing; this summary is for your convenience; the real discussion is in those chapters.

   Some parts of this evaluation will be hard to do objectively without assistance. For example, how will you find out whether your people feel psychologically safe? Another area where self-assessment may be challenging is whether you are open to changing your mind. So turn to your mentors, peers, and if appropriate in your cultural context, even subordinates who are confident enough to "tell truth to power."

2. Which of the areas in which you fall short are key to your current context?

3. The average person can tackle only one to three major changes in parallel. As such, you have to set priorities for development, or the challenge will seem overwhelming. The discussions in all the chapters roughly fall into three buckets: mindsets, behaviors and capabilities, and actions. Mindsets are typically in the first half of each chapter and actions in the second half. Behaviors are spread across.

Pick mutually reinforcing mindsets, behaviors and capabilities, and actions to work on. For example, if you choose to work on "Psychological safety is my responsibility" (mindset), then "Choose words carefully—words matter" (behavior) and "Reduce incomplete knowledge problem" (action) may be appropriate to work on. In contrast, "Communicate your strategic intent" should probably be deferred to another time as it won't contribute to improving psychological safety.

4. Add your choices to your personal leadership philosophy. Also decide on the conditions under which you will address elements you aren't prioritizing.

5. Develop a plan for how you will actually implement your new resolution. For example, if you need to improve your efforts on psychological safety significantly, recognize that people won't trust your motives initially. The first time you slip up (you will—these things are hard), they'll believe you're returning to your old norm. Appointing a "devil's advocate" who can publicly challenge you could help. What guidance would you give this person? How will you introduce the person to your people? Follow this logic with your other choices.

6. Discuss your decisions with a mentor who will give unvarnished feedback.

The Last Word ... Is "Why?"

Incorporating elements of the profile of a Producing Leader of Creativity into your personal leadership philosophy will make you a better leader of your organization's challenges and opportunities. There are two additional reasons why you should do so.

The first is that if you are smart, and I'm assuming you are, your country, and the world at large, need you to do so. Digital technologies are inextricably intertwined with today's challenges in very troubling ways. In addition to enhancing economic inequality, they enable organizations to pry into people's thoughts and shape their behaviors. In many countries, frustrated citizenry have turned to strongmen politicians only to see these technologies used by governments against them.[1]

Nobel laureate in Economics Joseph Stiglitz recently wrote an article that addressed America's current problems, which can easily be adapted to apply to most other countries.[2] Professor Stiglitz's message wasn't unique; others have recently made similar arguments. I picked his because in a very few sentences, he powerfully summarized the key issues:

> [T]he American economy is failing its citizens. . . . the fortunes of young Americans [are] more dependent on the income and education of their parents than elsewhere.
>
> . . . we can indeed channel the power of the market to serve society.
>
> America created the first truly middle-class society; now, a middle-class life is increasingly out of reach for its citizens.
>
> . . . we forgot that the true source of the wealth of a nation is the creativity and innovation of its people. . . . We confused the hard work of wealth creation with wealth-grabbing (or, as economists call it, rent-seeking), and too many of our talented young people followed the siren call of getting rich quickly.

Stiglitz's powerful message of what must change will come to naught unless business leaders embrace the model of a Producing Leadership of Creativity. In executive education and MBA classes, I don't mince words. "When they come with pitchforks, and they will," I say, "they'll come for you and me. We've created this world. They're paying the price." With that, I ask, "*What could you do?*" With the MBAs, I add that I'm all for their becoming wildly rich, provided they do so by developing businesses that improve the world, instead of the next addictive Candy Crush or Angry Birds app.

The second reason is an unprecedented opportunity. Though the twentieth century repeatedly sought "all hands on deck" in countries worldwide, it really meant only certain hands. The twenty-first century is asking—and will ask more fervently—for "all brains on deck." This time, there's the potential to truly mean "all." In a world that worships productivity, each individual is an easily replaceable cog. In the world that respects creativity, each individual merits attention.

How many groups lie ignored in the shadows despite their (potential for) amazing contributions? The digital world has the power to bring them to the forefront. Whether its power is used to do so is ultimately up to *you. You can't be a true leader if you aren't willing to remove the traditional blinders society puts over your eyes.* If you remove them, you'll have access to a talent pool no predecessor of yours ever did.

The Final "Why"?

Because someone needs to skeptically pursue big ideas to help humanity confront its challenges and opportunities. It might as well be you. Or better yet, many, many of you, since you need partners to collaborate. If you don't succeed, no specific harm will befall you. If you succeed, you'll improve the world—and may do very well for yourself.

Don't you like those odds?

Notes

Chapter 1

1. Ramchandran Jaikumar, "From Filing and Fitting to Flexible Manufacturing: A Study in the Evolution of Process Control," *Foundations and Trends(R) in Technology, Information and Operations Management* 1, no. 1 (2005): 1–120, https://ideas.repec.org/a/now/fnttom/0200000001.html.

2. Joyce Shaw Peterson, *American Automobile Workers, 1900–1933* (Albany: State University of New York Press, 1987).

3. Peter Nulty and Karen Nickel, "America's Toughest Bosses," *Fortune*, February 27, 1989, http://archive.fortune.com/magazines/fortune/fortune_archive/1989/02/27/71677/index.htm.

4. Brian Dumaine and Rosalind Berlin, "America's Toughest Bosses." *Fortune*, October 18, 1993, http://archive.fortune.com/magazines/fortune/fortune_archive/1993/10/18/78470/index.htm.

5. Daniel Goleman, "When the Boss Is Unbearable," *New York Times*, December 28, 1986, https://www.nytimes.com/1986/12/28/business/when-the-boss-is-unbearable.html.

6. In September 2018, Harvard Business School Press confirmed the case and video had been a part of its catalog and have long been withdrawn.

7. Hara Marano, "When the Boss Is a Bully," *Psychology Today*, September 1, 1995, https://www.psychologytoday.com/us/articles/199509/when-the-boss-is-bully.

8. Alex Hailey, *Wheels* (New York: Doubleday, 1971).

9. David Garvin, "Quality on the Line," *Harvard Business Review*, September 1983.

10. Kim B. Clark and Takahiro Fujimoto, *Product Development Performance: Strategy, Organization, and Management in the World Auto Industry* (Boston: Harvard Business Review Press, 1991).

11. Richard Pascale and Anthony Athos, *The Art of Japanese Management* (New York: Warner Books, 1982).

12. Matthias Weiss and Martin Hoegl, "The History of Teamwork's Societal Diffusion: A Multi-Method Review," *Small Group Research* 46, no. 6 (2015): 589–622, https://doi.org/DOI:10.1177/1046496415602778.

Chapter 2

1. Daniel M. Wegner and David J. Schneider, "The White Bear Story," *Psychological Inquiry* 14, no. 3–4 (2003): 326–29.

2. George Lakoff and Mark Johnson, *Metaphors We Live By* (Chicago: University of Chicago Press, 1980).

3. Richard Foster, *Innovation: The Attacker's Advantage* (New York: Summit Books, 1986).

4. Devendra Sahal, *Patterns of Technological Innovation* (Reading, MA: Addison-Wesley, 1981).

5. Joseph Bower and Clayton Christensen, "Disruptive Technologies: Catching the Wave," *Harvard Business Review*, January–February 1995.

6. Clayton Christensen, Michael Raynor, and Rory McDonald, "What Is Disruptive Innovation?," *Harvard Business Review*, December 2015.

7. KPMG, "The Pulse of Fintech 2018," updated February 2018, https://assets.kpmg/content/dam/kpmg/xx/pdf/2018/07/h1-2018-pulse-of-fintech.pdf, and "The Pulse of Fintech 2016," February 2017, https://assets.kpmg.content/dam/kpmg/xx/pdf/2017/02/pulse-of-fintech-q4-2016.pdf. United Nations, "Nominal GDP Data," https://unstats.un.org/unsd/snaama/Index.

8. Jessica Haywood, Patrick Mayock, Jan Freitag, Kwabena Akuffo Owoo, and Blase Fiorilla, *Airbnb & Hotel Performance*, STR (2017), http://www.str.com/Media/Default/Research/STR_AirbnbHotelPerformance.pdf.

9. Joyce E. Cutler, "Airbnb Paying Taxes in 275 Jurisdictions Worldwide," *Bloomberg News*, April 18, 2017, https://web.archive.org/web/20180827234758/https://www.bna.com/airbnb-paying-taxes-n57982086807.

10. W. Chan Kim and Renee Mauborgne, *Blue Ocean Strategy* (Boston: Harvard Business School Press, 2005).

11. Pui-Wing Tam, "How Silicon Valley Came to Be a Land of 'Bros,'" *New York Times*, February 5, 2018, https://www.nytimes.com/2018/02/05/technology/silicon-valley-brotopia-emily-chang.html.

12. Many industry executives and professionals and my own experiences provided inputs for this and the next section. Dr. Amitabha Chaudhuri, Chief Technology Officer of MedGenome, and formerly a research scientist at Genentech, Yale, and Harvard, provided extensive guidance. Any remaining errors are mine. For a complementary perspective, see Julian Birkinshaw, Ivanka Visnjic, and Simon Best, "Responding to a Potentially Disruptive Technology: How Big Pharma Embraced Biotechnology," *California Management Review* 60, no. 4 (2018): 74–100.

13. John C. Alexander and Daniel E. Salazar, "Modern Drug Discovery and Development," chap. 25 in *Clinical and Translational Science*, ed. David Robertson and Gordon H. Williams (Amsterdam: Elsevier, 2009), 361–380.

14. "The Human Genome Project Completion: Frequently Asked Questions," updated November 12, 2018, https://www.genome.gov/11006943/human-genome -project-completion-frequently-asked-questions.

15. Alexander and Salazar, "Modern Drug Discovery and Development."

16. Gunter Festel Alexander Schicker, and Roman Boutellier. "Performance Improvement in Pharmaceutical R&D through New Outsourcing Models," *Journal of Business Chemistry* 7, no. 2 (2010): 89–96.

17. Personal contact with executives and scientific professionals (2001–2010). This information is no longer subject to nondisclosure agreements.

18. Ramasastry Chandrashekhar and J. Robert Mitchell, "Boehringer Ingleheim: Leading Innovation," Ivey School of Business, 2014. Case.

19. Polina Bochukova and Donald A. Marchand, "Digital Transformation at Novartis to Improve Customer Engagement," IMD, 2014. Case.

20. James Frederick, "GSK Conference Explores Internet, Supply Chain," *Drug Store News*, June 17, 2002.

21. Edwin Lopez, "The Drug Supply Chain Security Act: A Progress Report," *Supply Chain Dive*, April 23, 2018, https://www.supplychaindive.com/news/Drug -Supply-Chain-Security-Act-progress-serialization-spotlight/521862.

22. "The DSCSA Pharmaceutical Serialization Deadline Looms," 2018, https:// www.datexcorp.com/the-dscsa-pharmaceutical-serialization-deadline-looms.

23. Laurie Sullivan, "FDA Approves RFID Tags for Humans," *InformationWeek*, October 14, 2004, https://www.informationweek.com/fda-approves-rfid-tags-for -humans/d/d-id/1027823.

24. Beth Bacheldor, "AMA Issues Ethics Code for RFID Chip Implants," *RFID Journal*, July 17, 2007, https://www.rfidjournal.com/articles/view?3487/2.

25. "FDA Approves Pill with Sensor That Digitally Tracks If Patients Have Ingested Their Medication." News release. November 13, 2017, https://www.fda .gov/newsevents/newsroom/pressannouncements/ucm584933.htm.

26. Jody Rosen, "The Knowledge, London's Legendary Taxi-Driver Test, Puts Up a Fight in the Age of GPS," *New York Times Style Magazine*, November 10, 2014.

27. "Hewlett Foundation Sponsors Prize to Improve Automated Scoring of Student Essays," news release, January 9, 2012, https://hewlett.org/newsroom/hewlett -foundation-sponsors-prize-to-improve-automated-scoring-of-student-essays.

28. Tovia Smith, "More States Opting to 'Robo-Grade' Student Essays by Computer," *National Public Radio*, June 30, 2018, https://www.npr.org/2018/06/30/624373367 /more-states-opting-to-robo-grade-student-essays-by-computer.

29. Julie Bort, "How IBM Watson Saved the Life of a Woman Dying from Cancer, Exec Says," *Business Insider*, December 7, 2016, https://www.business insider.com/how-ibm-watson-helped-cure-a-womans-cancer-2016-12.

30. Peerzada Abrar, "IBM's Supercomputer Helps Doctors to Fight Cancer," *The Hindu*, August 7, 2016, https://www.thehindu.com/business/IBM's-Supercomputer -helps-doctors-to-fight-cancer/article14556945.ece.

31. Yonghui Wu, Mike Schuster, Zhifeng Chen, Quoc V. Le, Mohammad Norouzi, Wolfgang Macherey, Maxim Krikun, et al., "Google's Neural Machine Translation System: Bridging the Gap between Human and Machine Translation," Cornell University, https://arxiv.org/abs/1609.08144#.

32. Fredric Lardinois, "Google Brings Offline Neural Machine Translations for 59 Languages to Its Translate App," *TechCrunch*, June 12, 2018, https://tech crunch.com/2018/06/12/google-brings-offline-neural-machine-translation-for -59-languages-to-its-translate-app.

33. Linda Carroll, "Google Translate Mostly Accurate in Test with Patient Instructions," *Reuters*, February 25, 2019, https://www.reuters.com/article/us -health-translations/google-translate-mostly-accurate-in-test-with-patient-inst- ructions-idUSKCN1QE2KB.

34. Elaine Khoon, "Assessing the Use of Google Translate for Spanish and Chi- nese Translations of Emergency Department Discharge Instructions," Research Letter, *Journal of the American Medical Association*, February 25, 2019, https:// jamanetwork.com/journals/jamainternalmedicine/article-abstract/2725080.

35. Tim O'Reilly, "What Will Our Lives Be Like as Cyborgs?," *The Atlantic*, October 27, 2017, https://www.theatlantic.com/technology/archive/2017/10 /cyborg-future-artificial-intelligence/543882.

36. Steve Viscelli, "Driverless? Autonomous Trucks and the Future of the American Trucker," September 2018, Center for Labor Research and Education, University of California, Berkeley, http://driverlessreport.org.

37. Alexis Madrigal, "Could Self-Driving Trucks Be Good for Truckers?," *The Atlantic*, February 1, 2018, https://www.theatlantic.com/technology/archive/2018/02/uber-says-its-self-driving-trucks-will-be-good-for-truckers/551879.

38. Rory McDonald and Suresh Kotha, "Boeing 787: Manufacturing a Dream," Harvard Business School, 2015. Case. These companies produced the hardware; several others created the software.

39. Suresh Kotha, "Boeing 787: The Dreamliner," Harvard Business School, 2008. Case.

40. Pierre Zahnd, "3D Printing in the Automotive Industry," *3D Printing Industry*, May 10, 2018, https://3dprintingindustry.com/news/3d-printing-automotive-industry-3-132584.

41. Brian Krassenstein, "20,000 3D Printed Parts Are Currently Used on Boeing Aircraft as Patent Filing Reveals Further Plans," March 7, 2015, https://3dprint.com/49489/boeing-3d-print.

42. Adam Jezard, "One-Quarter of Dubai's Buildings Will Be 3D Printed by 2025," *World Economic Forum*, May 15, 2018, https://www.weforum.org/agenda/2018/05/25-of-dubai-s-buildings-will-be-3d-printed-by-2025.

43. Jack Morley, "Architects: Here's the Problem with 3D-Printed Buildings," *Architizer*, undated, circa 2015, https://architizer.com/blog/practice/details/3d-printed-buildings-future-or-gimmick.

44. Jezard, "One-Quarter of Dubai's Buildings Will Be 3d Printed by 2025."

45. Matthew Shaer, "Soon, Your Doctor Could Print a Human Organ on Demand," *Smithsonian.com*, 2015, https://www.smithsonianmag.com/innovation/soon-doctor-print-human-organ-on-demand-180954951.

46. Chris Welch, "Google Just Gave a Stunning Demo of Assistant Making an Actual Phone Call," *The Verge*, May 8, 2018, https://www.theverge.com/2018/5/8/17332070/google-assistant-makes-phone-call-demo-duplex-io-2018.

47. R. Douglas Fields, "Wristband Lets the Brain Control a Computer with a Thought and a Twitch," *Scientific American*, March 27, 2018, https://www.scientificamerican.com/article/wristband-lets-the-brain-control-a-computer-with-a-thought-and-a-twitch.

48. Andreas Dür, Leonardo Baccini, and Manfred Elsig, "The Design of International Trade Agreements: Introducing a New Dataset," *The Review of International Organizations* 9, no. 3 (2014): 353–375.

49. "World Trade Statistical Review 2018," ed. World Trade Organization, 2018, https://www.wto.org/english/res_e/statis_e/wts2018_e/wts2018_e.pdf.

50. Dylan Martin, "GE Digital's Predix Now Supports Low-Latency, Offline Iot Deployments," *CRN*, November 1, 2018, https://www.crn.com/news/internet-of -things/ge-digital-s-predix-now-supports-low-latency-offline-iot-deployments.

51. Charles Rollet, "Ecuador's All-Seeing Eye Is Made in China," *Foreign Policy*, August 8, 2018, https://foreignpolicy.com/2018/08/09/ecuadors-all-seeing-eye-is -made-in-china.

52. Raquel Carvalho, "In Latin America, Big Brother China Is Watching You," *South China Morning Post*, December 21, 2018, https://www.scmp.com/week -asia/geopolitics/article/2178558/latin-america-big-brother-china-watching-you.

53. Kevin Vick, "The Right to Be Forgotten," undated, http://www.medialaw .org/component/k2/item/3994-the-right-to-be-forgotten.

54. Michal Kosinskia, David Stillwella, and Thore Graepel, "Private Traits and Attributes Are Predictable from Digital Records of Human Behavior," *Proceedings of the National Academy of Sciences* 110, no. 15 (2013), 5802–5805.

Chapter 3

1. "Q. Who First Originated the Term VUCA (Volatility, Uncertainty, Complexity and Ambiguity)?," US Army Heritage & Education Center website, updated May 7, 2019, http://usawc.libanswers.com/faq/84869.

2. Daniel Kahneman, *Thinking Fast and Slow* (New York: Farrar, Straus, and Giroux, 2013).

3. Cantor, Joanne. "Flooding Your Brain's Engine: How You Can Have Too Much of a Good Thing." *Psychology Today*, February 27, 2011, https://www .psychologytoday.com/us/blog/conquering-cyber-overload/201102/flooding -your-brain-s-engine-how-you-can-have-too-much-good. Also, Kahneman, *Thinking Fast and Slow*.

4. Ashadun Nobi, Sungmin Lee, Doo Hwan Kim, and Jae Woo Lee, "Correlation and Network Topologies in Global and Local Stock Indices," *Physics Letter A* 378, no. 34 (July 4, 2014): 2482–2489.

5. Bentian Li and Dechang Pi, "Analysis of Global Stock Index Data during Crisis Period via Complex Network Approach," *PLoS One* 13, e0200600, no. 7 (2018), https://www.ncbi.nlm.nih.gov/pmc/articles/PMC6051609/#sec011title.

6. Charles D. Brummitt and Teruyoshi Kobayashi, "Cascades in Multiplex Financial Networks with Debts of Different Seniority," *Physical Review E* 91, no. 062813 (June 24, 2015), https://journals.aps.org/pre/pdf/10.1103/PhysRevE.91.062813.

7. "Modeling How Contagion Spreads in a Financial Crisis," Columbia University website profiling work of Professor David Yao, Data Sciences Institute, Columbia University, updated May 6, 2015, https://datascience.columbia.edu/modeling-how-contagion-spreads-financial-crisis.

8. Andrew Sheng, *Financial Crisis and Global Governance: A Network Analysis*, http://siteresources.worldbank.org/EXTPREMNET/Resources/489960-133899 7241035/Growth_Commission_Working_Paper_67_Financial_Crisis_Global _Governance_Network_Analysis.pdf.

9. Camelia Minoiu, Chanhyun Kang, V.S. Subrahmanian, and Anamaria Berea, "Does Financial Connectedness Predict Crises?," *Quantitative Finance* 15, no. 4 (2015): 607–624.

10. Sigríður Benediktsdóttir, Gauti Bergþóruson Eggertsson, and Eggert Þórarinsson, *The Rise, Fall, and Resurrection of Iceland: A Postmortem Analysis of the 2008 Financial Crisis*, Brookings Papers on Economic Activity, https://www .brookings.edu/wp-content/uploads/2018/02/benediktsdottirtextfa17bpea .pdf.

11. J. R. Minkel, "The 2003 Northeast Blackout—Five Years Later," *Scientific American*, August 13, 2008, https://www.scientificamerican.com/article/2003-blackout -five-years-later.

Chapter 4

1. Gustavo Tavares, Filipe Sobral, Rafael Goldszmidt, and Felipe Araújo, "Opening the Implicit Leadership Theories' Black Box: An Experimental Approach with Conjoint Analysis," *Frontiers in Psychology*, February 7, 2018.

2. Pankaj Ghemawat and Herman Vantrappen, "How Global Is Your C-Suite?," *Sloan Management Review*, Summer 2015.

3. *Corporate Executive Board Global Labor Market Survey, Q3 2012; Russell Reynolds Associates, Asia Leadership Survey, 2012.*

4. J. Stewart Black and Alan Morrison, "The Japanese Global Leadership Challenge: What It Means for the Rest of the World," *Asia Pacific Business Review* 18, no. 4 (2012): 551–566.

5. Naoyuki Iwatani, Gordon Orr, and Brian Salsberg, "Japan's Globalization Imperative," *McKinsey Quarterly*, June 2011.

6. Justin Wolfers, "Fewer Women Run Big Companies Than Men Named John," *New York Times*, March 2, 2015, https://www.nytimes.com/2015/03/03/upshot /fewer-women-run-big-companies-than-men-named-john.html?_r=0.

7. *Heidrick & Struggles Route to the Top*, 2018, https://www.heidrick.com/Knowl edge-Center/Publication/Route_to_the_Top_2018.

8. Daniel Goleman, "Leadership That Gets Results," *Harvard Business Review*, March–April 2000.

9. Weinberger, David, "'Powering Down' Leadership in the U.S. Army," *Harvard Business Review*, November 2010.

10. Devin Hargrove and Sim Sitkin. "Next Generation Leadership Development in a Changing and Complex Environment: An Interview with General Martin E. Dempsey," *Academy of Management Learning & Education* 10, no. 3 (2011): 528–533.

11. Stanley McChrystal, Tantum Collins, David Silverman, and Chris Fussell, *New Rules of Engagement for a Complex World* (New York: Portfolio/Penguin, 2015).

12. David J. Armstrong and Paul Cole, "Managing Distances and Differences in Geographically Distributed Work Groups," in *Distributed Work*, ed. Pamela Hinds and Sara Kiesler (Cambridge: MIT Press, 2002), 167–189.

13. Cristina Gibson, Laura Huang, Bradley Kirkman, and Debra L. Shapiro, "Where Global and Virtual Meet: The Value of Examining the Intersection of These Elements in Twenty-First-Century Teams," *Annual Review of Organizational Psychology and Organizational Behavior*, no. 1 (2014): 217–244.

14. Faaiza Rashid and Amy Edmondson, "Risky Trust: How Teams Build Trust Despite High Risk," *Rotman Magazine*, Spring 2012.

15. Catherine Cramton and Pamela Hinds, "An Embedded Model of Cultural Adaptation in Global Teams," *Organization Science* 25, no. 4 (2014): 1056–1081.

16. Amit Mukherjee, "The Effective Management of Organizational Learning and Process Control in the Factory," DBA thesis, Harvard University, 1992.

17. Karen Christensen, "Thought Leader Interview: Amy Edmondson," *Rotman Magazine*, Winter, 2013.

Chapter 5

1. Samuel Palmisano, "The Global Enterprise," *Foreign Affairs*, October 14, 2016, https://www.foreignaffairs.com/articles/2016-10-14/global-enterprise.

2. David A. Thomas, "Diversity as Strategy," *Harvard Business Review*, September 2004.

3. Rosabeth Moss Kanter, "IBM in the 21st Century: The Coming of the Globally Integrated Enterprise," Harvard Business School, 2009. Case.

4. "Developing Global Leadership," IBM white paper, February 2010, https://www.ibm.com/downloads/cas/K7EWX39G.

5. "The IBMer," IBM Corporate Responsibility Report, 2012, https://www.ibm.com/ibm/responsibility/2012/bin/downloads/ibm_crr2012_the_ibmer.pdf.

6. Douglas Ready and M. Ellen Peebles, "Developing the Next Generation of Enterprise Leaders," *Sloan Management Review*, Fall 2015.

7. Personal conversations with current and former Unilever executives in Asia.

8. "East Asians Less Likely to Occupy Leadership Roles Than South Asians at US Companies, Asia Society Survey Finds," Asia Society, updated May 25, 2017, https://asiasociety.org/media/east-asians-less-likely-occupy-leadership-roles-south-asians-us-companies-asia-society-survey-.

9. Winter Nie, Daphne Xiao, and Jean-Louis Barsoux. "Rethinking the East Asian Leadership Gap," *Sloan Management Review*, Summer 2017.

10. "Leadership Development & Performance Management," archived version of J&J website from April 3, 2017, https://web.archive.org/web20170403222638/http://www.jnj.com/caring/citizenship-sustainability/leadership-development-and-performance-management.

11. "Change Waits for No One—We Shape the Future," Daimler website, https://www.daimler.com/career/about-us/culture-benefits/leadership-2020/.

12. Hope King, "Salesforce CEO: I Didn't Focus on Hiring Women Then. But I Am Now," *CNN Business*, June 12, 2015, https://money.cnn.com/2015/06/12/technology/salesforce-ceo-women-equal-pay/index.html.

13. Niclas Erhardt, James Werbel, and Charles Shrader, "Board of Director Diversity and Firm Financial Performance," *Corporate Governance: An International Review* 11 (2003): 102–111.

14. Kevin Campbell and Antonio Mínguez-Vera, "Gender Diversity in the Board-room and Firm Financial Performance," *Journal of Business Ethics* 83, no. 3 (2008): 435–51.

15. Mijntje Lückerath-Rovers, "Women on Boards and Firm Performance," *Journal of Management & Governance* 17, no. 2 (May 2013): 491–509.

16. Larelle Chapple and Jacquelyn Humphrey, "Does Board Gender Diversity Have a Financial Impact? Evidence Using Stock Portfolio Performance," *Journal of Business Ethics* 122, no. 4 (2014): 709–723.

17. Stephen Bear, Noushi Rahman, and Corrine Post. "The Impact of Board Diversity and Gender Composition on Corporate Social Responsibility and Firm Reputation," *Journal of Business Ethics* 97, no. 2 (2010): 207–221.

18. Pnina Shachaf, "Cultural Diversity and Information and Communication Technology Impacts on Global Virtual Teams: An Exploratory Study," *Information and Management* 45, no. 2 (March 2008): 131–142.

19. Vivian Hunt, Sara Prince, Sundiatu Dixon-Fyle, and Lareina Yee, "Delivering through Diversity," *McKinsey Global Institute*, January 2018, https://www.mckinsey.com/~/media/mckinsey/business%20functions/organization/our%20insights/delivering%20through%20diversity/delivering-through-diversity_full-report.ashx.

20. Iwatani et al., "Japan's Globalization Imperative."

21. Nitasha Tiku, "Google Ends Forced Arbitration after Employee Protest," *Wired*, February 21, 2019, https://www.wired.com/story/google-ends-forced-arbitration-after-employee-protest.

22. Jena McGregor, "Google Has Fired the Employee behind That Controversial Diversity Manifesto," *Washington Post*, August 7, 2017, https://www.washingtonpost.com/news/on-leadership/wp/2017/08/07/a-googlers-manifesto-is-the-hr-departments-worst-nightmare/?utm_term=.5757a67c700c.

23. Clive Thompson, "The Secret History of Women in Coding," *New York Times Magazine*, February 13, 2019, https://www.nytimes.com/2019/02/13/magazine/women-coding-computer-programming.html?searchResultPosition=1.

24. US Department of Labor Statistics, https://www.dol.gov/wb/stats/NEWSTATS/latest.htm#LFPRates.

25. National Center for Education Statistics, https://nces.ed.gov/programs/digest/d12/tables/dt12_349.asp.

26. Clive Thompson, "The Secret History of Women in Coding."

27. Jane Margolis and Allan Fisher, *Unlocking the Clubhouse: Women in Computing* (Cambridge: MIT Press, 2002).

28. Emma Featherstone, "Why Women in Stem May Be Better Off Working in India and Latin America," *Guardian*, June 24, 2015, https://www.theguardian.com/guardian-professional/2015/jun/24/why-women-in-stem-may-be-better-off-working-in-india-and-latin-america.

29. Peerzada Abrar, "Rise, and Rise, of the Women Who Code," *The Hindu*, March 28, 2016, https://www.thehindu.com/business/Industry/rise-and-rise-of-the-women-who-code/article8402484.ece.

30. Clive Thompson, "The Secret History of Women in Coding."

31. Yonatan Zunger, "So, about This Googler's Manifesto," August 5, 2017, https://medium.com/@yonatanzunger/so-about-this-googlers-manifesto-1e3773ed1788.

32. Jonathan Woetzel, Anu Madgavkar, Kweilin Ellingrud, Eric Labaye, Sandrine Devillard, Eric Kutcher, James Manyika, et al., "How Advancing Women's Equality Can Add $12 Trillion to Global Growth," *McKinsey Global Institute*, September 2015, https://www.mckinsey.com/featured-insights/employment-and-growth/how-advancing-womens-equality-can-add-12-trillion-to-global-growth.

33. Vivian Hunt et al., "Delivering through Diversity."

34. Tsedal Neeley, "Global Teams That Work," *Harvard Business Review*, October 2015.

35. Stephen Turban, Dan Wu, and Letian (LT) Zhang, "When Gender Diversity Makes Firms More Productive," *Harvard Business Review*, February 11, 2019.

36. Vijay Govindrajan and Anil K. Gupta, "Building an Effective Global Business Team," *Sloan Management Review*, Summer 2001.

37. Ara Norenzayan, "Explaining Human Behavioral Diversity," *Science* 332, no. 6033 (May 27 2011): 1041–1042.

38. Jesse Harrington and Michele Gelfand, "Tightness-Looseness across the 50 United States," *Proceedings of the National Academy of Sciences* 111, no. 22 (2014): 7990–7995.

39. Janet Bennett and Milton Bennett, *Developing Intercultural Sensitivity: An Integrated Approach to Global and Domestic Diversity*, http://www.diversitycollegium.org/pdf2001/2001Bennettspaper.pdf.

40. Personal contact with executives and scientific professionals (2001–2010). This information is no longer subject to non-disclosure agreements.

41. Peter Kuhn and Marie-Claire Villeval, "Are Women More Attracted to Cooperation Than Men?," *The Economic Journal* 125, no. 582 (February 2015): 115–140.

42. Eduardo Araújo, Nuno Araújo, André Moreira, Hans Herrmann, and José Andrade Jr. "Gender Differences in Scientific Collaborations: Women Are More Egalitarian Than Men," *PLoS One* 12, no. 5 (2017): w0176791.

43. Bill Taylor, "The Logic of Global Business: An Interview with ABB's Percy Barnevik," *Harvard Business Review*, March–April 1991.

44. W. Chan Kim and Renee Mauborgne, "Procedural Justice, Attitudes, and Subsidiary Top Management Compliance with Multinationals' Corporate Strategic Decisions," *Academy of Management Journal* 36, no. 3 (1993): 502–526.

45. Tsedal Neeley and Thomas J. DeLong, "Managing a Global Team: Greg James at Sun Microsystems, Inc. (a)," Harvard Business School, November 1, 2009. Case.

46. Kahneman, *Thinking Fast and Slow*.

47. Jon J. Nordby, "Can We Believe What We See, If We See What We Believe?—Expert Disagreement," *Journal of Forensic Sciences* 37, no. 4 (1992): 1115–1124.

48. Donna Chrobot-Mason and Nicholas Aramovich, "The Psychological Benefits of Creating an Affirming Climate for Workplace Diversity," *Group & Organization Management* 38, no. 6 (2013): 659–689.

49. Eric Kearney, Diether Gebert, and Sven C. Voelpel, "When and How Diversity Benefits Teams: The Importance of Team Members' Need for Cognition," *Academy of Management Journal* 52, no. 3 (2009): 581–598.

50. Scott Page, "Making the Difference: Applying a Logic of Diversity," *Academy of Management Perspectives* 21, no. 4 (November 1, 2007).

51. Anthony G. Greenwald and Linda Hamilton Krieger, "Implicit Bias: Scientific Foundations," *California Law Review* 94, no. 4 (2006): 945–967.

52. Adil H. Haider, Janel Sexton, and N. Sriram, "Association of Unconscious Race and Social Class Bias with Vignette-Based Clinical Assessments by Medical Students," *Journal of the American Medical Association* 306, no. 9 (2011): 942–951.

53. Dana Kanze, Laura Huang, Mark A. Conley, and E. Tory Higgins, "We Ask Men to Win and Women Not to Lose: Closing the Gender Gap in Startup Funding," *Academy of Management Journal* 61, no. 2 (2018): 586–614.

54. Victoria Brescoll and Eric Uhlmann, "Can an Angry Woman Get Ahead? Status Conferral, Gender, and Expression of Emotion in the Workplace," *Psychological Science* 19, no. 3 (2008): 268–275.

55. Seval Gündemir, Astrid C. Homan, Carsten K. W. de Dreu, and Mark van Vugt, "Think Leader, Think White? Capturing and Weakening an Implicit Pro-White Leadership Bias," *Plos One*, January 8, 2014.

56. Michael Grothaus, "How 'Blind Recruitment' Works and Why You Should Consider It," *Fast Company*, March 14, 2016, https://www.fastcompany .com/3057631/how-blind-recruitment-works-and-why-you-should-consider.

Chapter 6

1. Morton Hansen, "IDEO CEO Tim Brown: T-Shaped Stars: The Backbone of IDEO's Collaborative Culture," *Chief Executive Magazine*, January 21, 2010, https://chiefexecutive.net/ideo-ceo-tim-brown-t-shaped-stars-the-backbone-of -ideoaes-collaborative-culture_trashed.

2. Warren Benis and Robert Thomas, "Crucibles of Leadership," *Harvard Business Review*, September 2002.

3. Boris Groysberg, L. Kevin Kelly, and Bryan MacDonald, "The New Path to the C-Suite," *Harvard Business Review*, March 2011.

4. Michael Watkins, "When Managers Become Leaders," *Harvard Business Review*, June 2012.

5. Daniel Kahneman, *Thinking Fast and Slow* (New York: Farrar, Straus, and Giroux, 2013). My few sentences capture but simplify the essence of Kahneman's brilliant and highly readable work.

6. Amos Tversky and Daniel Kahneman, "Availability: A Heuristic for Judging Frequency and Probability," *Cognitive Psychology* 5 (1973): 207–232.

7. Gabrielle Hogan, "People Who Speak Multiple Languages Make the Best Employees, for One Big Reason," *Quartz*, March 9, 2017, https://qz.com/927660/ people-who-speak-multiple-languages-make-the-best-employees-for-one-big -reason.

8. Boaz Keysar, Sayuri L. Hayakawa, and Sun Gyu An, "The Foreign-Language Effect: Thinking in a Foreign Tongue Reduces Decision Biases," *Psychological Science* 23, no. 6 (June 2012): 661–668.

9. Robert Root-Bernstein, Lindsay B. Allen, Leighanna Beach, Ragini Bhadula, Justin Fast, Chelsea Hosey, Benjamin G. Kremkow, et al., "Arts Foster Scientific Success: Avocations of Nobel, National Academy, Royal Society, and Sigma Xi Members," *Journal of Psychology of Science and Technology* 1, no. 2 (October 2008): 51–63.

10. Cramton and Hinds, "An Embedded Model of Cultural Adaptation in Global Teams."

11. Gary Hamel and C. K. Prahalad, "Strategic Intent," *Harvard Business Review*, July–August 2005.

Chapter 7

1. Graham Allison and Philip Zelikow, *Essence of Decision: Explaining the Cuban Missile Crisis,* 2nd ed. (New York: Longman, 1999).

2. Kim B. Clark, "The Interaction of Design Hierarchies and Market Concepts in Technological Evolution," *Research Policy* 14, no. 5 (October 1985): 235–251.

3. Li-Ling Hsu and Minder Chen, "Impacts of ERP Systems on the Integrated-Interaction Performance of Manufacturing and Marketing," *Industrial Management & Data Systems* 104, no. 1: 42–55.

4. James Thompson, *Organizations in Action* (New York: McGraw-Hill, 1967).

5. Praveen Pinjani and Prashant Palvia, "Trust and Knowledge Sharing in Diverse Global Teams," *Information & Management* 50, no. 4 (2013): 144–153.

6. McDonald and Kotha, "Boeing 787: Manufacturing a Dream."

7. A. Elangovan and Debra Shapiro, "Betrayal of Trust in Organizations," *Academy of Management Review* 23, no. 3 (1998): 547–566.

8. Faaiza Rashid, and Amy Edmondson, "Risky Trust: How Teams Build Trust Despite High Risk," *Rotman Magazine*, Spring 2012.

9. Brad Crisp and Sirkka Jarvenpaa, "Swift Trust in Global Virtual Teams," *Journal of Personnel Psychology* 12, no. 1 (2013): 45–56.

10. Armstrong and Cole, "Managing Distances and Differences in Geographically Distributed Work Groups."

11. Ranjay Gulati, Franz Wohlgezogen, and Pavel Zhelyazkov, "The Two Facets of Collaboration: Cooperation and Coordination in Strategic Alliances," *The Academy of Management Annals* 6, no. 1 (2012): 531–583, https://doi.org/10.108 0/19416520.2012.691646.

12. Robert Axelrod, *The Evolution of Cooperation* (New York: Basic Books, 1984).

13. Julia Boorstin, "The Best Advice I Ever Got," *Fortune*, March 21, 2005.

14. Gender wage gap, OECD data 2016, https://data.oecd.org/earnwage/gender -wage-gap.htm.

15. Elahe Izadi, "Michelle Williams Got Paid Way Less Than Her Male Co-Star. It's a Sad Hollywood Tradition," *Washington Post*, January 10, 2018, https://www.washingtonpost.com/news/arts-and-entertainment/wp/2018/01/10/michelle-williams-got-paid-way-less-than-her-male-co-star-its-a-sad-hollywood-tradition/?utm_term=.1bb1684bc314.

16. Morten Hansen, "How John Chambers Learned to Collaborate at Cisco," *Harvard Business Review*, March 4, 2010.

17. Herminia Ibarra and Morten Hansen, "Are You a Collaborative Leader?," *Harvard Business Review*, July 1, 2011.

18. Elangovan and Shapiro, "Betrayal of Trust in Organizations."

19. Rashid and Edmondson, "Risky Trust."

20. Pinjani and Palvia, "Trust and Knowledge Sharing in Diverse Global Teams."

21. Rashid and Edmondson, "Risky Trust."

22. Crisp and Jarvenpaa, "Swift Trust in Global Virtual Teams."

23. M. Travis Maynard, John E. Mathieu, Tammy L. Rapp, and Lucy L. Gilson, "Something(s) Old and Something(s) New: Modelling Drivers of Global Virtual Team Effectiveness," *Journal of Organizational Behavior* 3, no. 3 (April 12, 2012): 342–365.

24. Eric Jackson, "The Seven Habits of Spectacularly Unsuccessful Executives," *Forbes*, January 2, 2012, https://www.forbes.com/sites/ericjackson/2012/01/02/the-seven-habits-of-spectacularly-unsuccessful-executives/#4865834e516b.

25. Amy Edmondson and Diana M. Smith, "Too Hot to Handle? How to Manage Relationship Conflicts," *California Management Review* 49, no. 1 (Fall 2006): 6–31.

26. Roderick Swaab, Katherine Phillips, and Michael Schaerera, "Secret Conversation Opportunities Facilitate Minority Influence in Virtual Groups: The Influence on Majority Power, Information Processing, and Decision Quality," *Organizational Behavior and Human Decision Processes* 133 (2016): 17–32.

27. Amy Edmondson, "Strategies for Learning from Failure," *Harvard Business Review*, April 2011.

28. Karen Christensen, "Thought Leader Interview: Amy Edmondson."

29. Polly Rizova, "Are You Networked for Innovation?," *Sloan Management Review*, Spring 2006.

30. Rob Cross, Peter Gray, Shirley Cunningham, Mark Showers, and Robert J. Thomas, "The Collaborative Organization: How to Make Employee Networks Really Work," *Sloan Management Review*, Fall, October 1, 2010.

Chapter 8

1. Ron Ashkenas, "It's Time to Rethink Continuous Improvement," *Harvard Business Review*, May 8, 2012.

2. Charalampos Mainemelis, Ronit Kark, and Olga Epitropaki, "Creative Leadership: A Multi-Concept Conceptualization," *The Academy of Management Annals* 9, no. 1 (2015): 393–482.

3. IBM, "IBM 2010 Global CEO Study: Creativity Selected as Most Crucial Factor for Future Success," news release, 2010, https://www-03.ibm.com/press /us/en/pressrelease/31670.wss.

4. Forrester Consulting, "The Creative Dividend: How Creativity Impacts Business Results," August 2014, https://landing.adobe.com/dam/downloads/white papers/55563.en.creative-dividends.pdf. This Adobe-sponsored study by Forrester Research of "managers and above who influence creative design software decisions" found fostering creativity paid off in revenue grown, market share, and employee satisfaction, but "61% of companies do not see their companies as creative."

5. Ronit Kark, Ella Miron-Spektor, Roni Gorsky, and Anat Kaplun, *Two Roads Diverge in a Yellow Wood: The Effect of Exploration and Exploitation on Creativity and Leadership Development*, working paper, Bar-Ilan University, 2014.

6. Jennifer Mueller, Jack Goncalo, and Dishan Kamdar, "Recognizing Creative Leadership: Can Creative Idea Expression Negatively Relate to Perceptions of Leadership Potential?," *Journal of Experimental Social Psychology* 47, no. 2 (2011): 494–498.

7. Jennifer Mueller, Shimul Melwani, and Jack Goncalo, "The Bias against Creativity: Why People Desire but Reject Creative Ideas," *Psychological Science* 23, no. 1 (2012): 13–17.

8. Robert Solow, "We'd Better Watch Out," *New York Times Book Review*, July 12, 1987.

9. Daron Acemoglu, David Autor, David Dorn, Gordon H. Hanson, and Brendan Price, "Return of the Solow Paradox? IT, Productivity, and Employment in US Manufacturing," *American Economic Review: Papers & Proceedings* 104, no. 5 (2014): 394–399.

10. "23 Economic Experts Weigh In: Why Is Productivity Growth So Low?," April 20, 2017, https://www.focus-economics.com/blog/why-is-productivity -growth-so-low-23-economic-experts-weigh-in.

11. Michael Mumford, Shane Connelly, and Blaine Gaddis, "How Creative Leaders Think: Experimental Findings and Cases," *Leadership Quarterly* 14, no. 4–5 (2003): 411–432.

12. Roni Reiter-Palmon and Jody J. Illies, "Leadership and Creativity: Understanding Leadership from a Creative Problem-Solving Perspective," *The Leadership Quarterly* 15, no. 1 (February 1 2004): 55–77.

13. Mumford et al., "How Creative Leaders Think: Experimental Findings and Cases."

14. Steve Wheelwright and Kim B. Clark, *Revolutionizing Product Development* (New York: Free Press, 2011).

15. Rizova, "Are You Networked for Innovation?"

16. Ed Catmull, "Inside the Pixar Braintrust," *Fast Company*, March 1, 2014, https://www.fastcompany.com/3027135/inside-the-pixar-braintrust.

17. Andy Boynton and Bill Fischer, *Virtuoso Teams: Lessons from Teams That Changed Their Worlds* (New York: FT Press, 2005).

18. "The Deep Dive with IDEO," ABC News Nightline, 1999, https://www.you tube.com/playlist?list=PL65FF22BBC5A7A59C.

19. Rob Goffee and Gareth Jones, "Leading Clever People," March 2007.

20. Mark Marotto, Johan Roos, and Bart Victor, "Collective Virtuosity in Organizations: A Study of Peak Performance in an Orchestra," *Journal of Management Studies* 44, no. 3 (March 26, 2007): 388–413.

21. Barbara Slavich, Rossella Cappetta, and Severino Salvemini, "Creativity and the Reproduction of Cultural Products: The Experience of Italian Haute Cuisine Chefs," *International Journal of Arts Management* 16, no. 2 (Winter 2014): 29–41, 70–71.

22. Marotto et al., "Collective Virtuosity in Organizations."

23. James G. (Jerry) Hunt, George E. Stellutob, and Robert Hooijberg, "Toward New-Wave Organization Creativity: Beyond Romance and Analogy in the Relationship between Orchestra-Conductor Leadership and Musician Creativity," *The Leadership Quarterly* 15, no. 1 (2004): 145–162.

24. Morag Barrett, "4 Leadership Lessons from Orchestra Conductors," *Entrepreneur*, May 20, 2015, https://entrepreneur.com/article/246194.

25. Robert Mirakian, "A Graduate Curriculum for Orchestral Conductors," unpublished doctoral thesis, Indian University Jacobs School of Music, May 2015, https://scholarworks.iu.edu/dspace/bitstream/handle/2022/19825/Mirakian,%20 Robert%20(DM%20Instrumental%20Conducting).pdf;jsessionid=3C8DEF4506F3 A06F0E616FBB84C9813C?sequence=1.

Pierfrancesco Bellini, Fabrizio Fioravanti, and Paolo Nesi, "Managing Music in Orchestras," *Computer* 32, no. 9 (September 1999): 26–34.

"How to Become a Music Conductor: Education and Career Roadmap," Study. com, https://study.com/articles/How_to_Become_a_Music_Conductor_Education _and_Career_Roadmap.html.

Sumarga Suanda, "How a Conductor Prepares for an Orchestral Performance," October 7, 2015, https://scholarblogs.emory.edu/cmbc/2015/10/07/how-a-con ductor-prepares-for-an-orchestral-performance/.

Courtney Lewis, "Conducting Electricity: Hearing the Entire Symphony Takes Years of Practice," *Florida Times-Union*, posted April 29, 2018; updated April 30, 2018, https://www.jacksonville.com/entertainmentlife/20180429/conducting -electricity-hearing-entire-symphony-takes-years-of-practice.

26. Krista Hyde, Jason Lerch, Andrea Norton, Marie Forgeard, Ellen Winner, Alan C. Evans, and Gottfried Schlaug, "Musical Training Shapes Structural Brain Development," *Journal of Neuroscience* 29, no. 10 (2009): 3019–3025.

27. Clemens Wöllner and Andrea Halpren, "Attentional Flexibility and Memory Capacity in Conductors and Pianists," *Attention, Perception, & Psychophysics* 78 (2016): 198–208.

28. Marotto et al., "Collective Virtuosity in Organizations."

29. Helen Rosner, "One Year of #Metoo: A Modest Proposal to Help Combat Sexual Harassment in the Restaurant Industry," *New Yorker*, October 10, 2018, https://www.newyorker.com/culture/annals-of-gastronomy/one-year-of-metoo -a-modest-proposal-to-help-dismantle-the-restaurant-industrys-culture-of -sexual-harassment.

30. Nicholas Gill, "Culinary Women Serve Up Their Own #Metoo Moment in Sweden," *Guardian*, March 2, 2018, https://www.theguardian.com/lifeandstyle/2018 /mar/02/culinary-women-serve-up-their-own-metoo-moment-in-sweden.

31. Mainemelis et al., "Creative Leadership."

32. J. Keith Murnighan and Donald E. Conlon, "The Dynamics of Intense Work Groups: A Study of British String Quartets," *Administrative Science Quarterly* 36, no. 2 (1991): 165–186.

33. Mainemelis et al., "Creative Leadership."

34. Mainemelis et al., "Creative Leadership."

35. Kelly Lindberg, "The Imaginer," *Continuum: The Magazine of the University of Utah*, Spring 2013, https://continuum.utah.edu/features/the-imaginer.

36. Nic Vargus, "Pixar's Ed Catmull on Taking Risks and Checking Your Ego," September 7, 2018, https://slackhq.com/pixars-ed-catmull-on-taking-risks-and-checking-your-ego.

37. Vargus, "Pixar's Ed Catmull on Taking Risks and Checking Your Ego."

38. Ed Catmull, "How Pixar Fosters Creativity," *Harvard Business Review*, September 2008.

39. "Conducting Successful Gate Meetings," Project Management.com, updated February 13, 2017, https://project-management.com/conducting-successful-gate-meetings.

40. Catmull, "How Pixar Fosters Creativity."

41. Catmull, "How Pixar Fosters Creativity."

42. Mohammadreza Hojat, Michael J. Vergare, Kaye Maxwell, George Brainard, Steven K. Herrine, Gerald A. Isenberg, Jon Veloski, and Joseph S. Gonnella, "The Devil Is in the Third Year: A Longitudinal Study of Erosion of Empathy in Medical School," *Academic Medicine* 84, no. 9 (September 2009): 1182–1191.

43. Helen Riess, "The Science of Empathy," *Journal of Patient Experience* 4, no. 2 (June 2017): 74–77.

44. Beth Howard, "Kindness in the Curriculum," AAMC News, September 18, 2018, https://news.aamc.org/medical-education/article/putting-kindness-curriculum.

45. Soloman Asch, "Studies of Independence and Conformity: I. A Minority of One against a Unanimous Majority," *Psychological Monographs: General and Applied* 70, no. 9 (1956): 1–70.

46. James E. Ryan, *Wait, What? And Life's Other Essential Questions* (New York: HarperCollins, 2017).

47. Andrew Hargaden and Beth Bechky, "When Collections of Creatives Become Creative Collectives: A Field Study of Problem Solving at Work," *Organization Science* 17, no. 4 (2006): 484–500.

48. Catmull, "How Pixar Fosters Creativity."

49. Hargaden and Bechky, "When Collections of Creatives Become Creative Collectives."

50. Amit Mukherjee, "The Effective Management of Organizational Learning and Process Control in the Factory," unpublished doctoral thesis, Harvard University, 1992.

51. Cristina B. Gibson and Jennifer L. Gibbs, "Unpacking the Concept of Virtuality: The Effects of Geographic Dispersion, Electronic Dependence, Dynamic Structure, and National Diversity on Team Innovation," *Administrative Sciences Quarterly* 51, no. 3 (September 2006): 451–495.

52. Michael A. Roberto, *Unlocking Creativity: How to Solve Any Problem and Make the Best Decisions by Shifting Creative Mindsets* (Hoboken, NJ: Wiley, 2019). Roberto calls this the "benchmarking mindset." A related issue is the "prediction mindset," which only pursues opportunities that are highly likely to succeed.

53. Roberto, *Unlocking Creativity*.

54. Navi Radjou, Jaideep Prabhu, and Simone Ahuja. *Jugaad Innovation: Think Frugal, Be Flexible, Generate Breakthrough Growth* (Hoboken, NJ: Wiley, 2012).

55. Catmull, "How Pixar Fosters Creativity."

56. Jennifer J. Deal, "Welcome to the 72-Hour Work Week," *Harvard Business Review*, September 13, 2013.

57. Jane Margolis and Allan Fisher, *Unlocking the Clubhouse: Women in Computing* (Cambridge: MIT Press, 2002).

58. Clive Thompson, "The Secret History of Women in Coding."

59. Albert Einstein, *Cosmic Religion: With Other Opinions and Aphorisms.* Covici-Friede, 1931.

60. Max Nisen, "Why GE Had to Kill Its Annual Performance Reviews after More Than Three Decades," *Quartz*, August 13, 2016, https://qz.com/428813/ge-performance-review-strategy-shift.

61. Peter Cappelli and Anna Tavis, "The Performance Management Revolution," *Harvard Business Review*, October 2016. This article also has a nice summary of the evolution of performance appraisals in the United States, starting with the efforts of the US Army during WW I.

62. Kris Duggan, "Why the Annual Performance Review Is Going Extinct," *Fast Company*, October 20, 2015.

63. Max Nisen, "Why GE Had to Kill Its Annual Performance Reviews after More Than Three Decades," *Quartz*, August 13, 2016, https://qz.com/428813/ge -performance-review-strategy-shift.

64. Marcus Buckingham and Ashley Goodall, "Reinventing Performance Management," *Harvard Business Review*, April 2015.

65. Amit Mukherjee, "It May Be Time to Get Rid of 'Smart' Management," *Forbes*, January 12, 2016, https://www.forbes.com/sites/forbesleadershipforum /2016/01/12/it-may-be-time-to-get-rid-of-smart-management/#3f7632273c07.

66. Hope King, "Salesforce CEO: I Didn't Focus on Hiring Women Then. But I Am Now," *CNN Business*, June 12, 2015, https://money.cnn.com/2015/06/12 /technology/salesforce-ceo-women-equal-pay/index.html.

Chapter 9

1. "Welcome to the Crisis Era: Are You Ready?," PriceWaterhouseCooper, 2017, https://www.pwc.com/gx/en/ceo-agenda/pulse/crisis.html.

2. Matthew Rosenberg, Nicholas Confessore, and Carole Cadwalladr, "How Trump Consultants Exploited the Facebook Data of Millions," *New York Times*, March 17, 2018, https://www.nytimes.com/2018/03/17/us/politics/cambridge -analytica-trump-campaign.html#click=https://t.co/UAg1Q5t1BG.

3. Carole Cadwalladr and Emma Graham-Harrison, "Revealed: 50 Million Facebook Profiles Harvested for Cambridge Analytica in Major Data Breach," *Guardian*, March 17, 2018, https://www.theguardian.com/news/2018/mar/17/cambridge -analytica-facebook-influence-us-election.

4. Basil Peters, "Venture Capital Exit Times," 2009, http://www.angelblog.net /Venture_Capital_Exit_Times.html.

5. Kate Clark, "VC Investment-to-Exit Ratio in the US at Record High," *PitchBook*, updated July 28, 2017, https://pitchbook.com/news/articles/vc-investment -to-exit-ratio-in-the-us-at-record-high.

6. Tia Ghose, "What Facebook Addiction Looks Like in the Brain," *LiveScience*, updated January 27, 2015, https://www.livescience.com/49585-facebook-addiction -viewed-brain.html.

7. David Lazarus, "Facebook Says You 'Own' All the Data You Post. Not Even Close, Say Privacy Experts," *Los Angeles Times*, March 19, 2018, https://www .latimes.com/business/lazarus/la-fi-lazarus-facebook-cambridge-analytica -privacy-20180320-story.html.

8. Aleksandra Korolova, "Facebook's Illusion of Control over Location-Related Ad Targeting," *Medium*, December 18, 2018, https://medium.com/@korolova/face books-illusion-of-control-over-location-related-ad-targeting-de7f865aee78.

9. Elizabeth Dwoskin and Craig Timberg, "Facebook Discussed Using People's Data as a Bargaining Chip, Emails and Court Filings Suggest," *Washington Post*, November 30, 2018, https://www.washingtonpost.com/technology/2018/11/30 /facebook-used-peoples-data-bargaining-chip-emails-court-filings-suggest/?utm _term=.9796f3856688.

10. Jennifer Valentino-DeVries, Natasha Singer, Michael H. Keller, and Aaron Krolik, "Your Apps Know Where You Were Last Night, and They're Not Keeping It Secret," *New York Times*, December 10, 2018, https://www.nytimes.com/interac tive/2018/12/10/business/location-data-privacy-apps.html.

11. Ananya Bhattacharya, "Facebook Patent: Your Friends Could Help You Get a Loan—Or Not," *CNN Business*, August 4, 2015, https://money.cnn.com /2015/08/04/technology/facebook-loan-patent.

12. Casey Newton, "It's Time to Regulate Tech Platforms with Laws, Not Fines," *The Verge*, July 30, 2019, https://www.theverge.com/interface/2019/7/30 /20746427/facebook-ftc-settlement-congress-privacy-law.

13. Alina Selyukh, "Section 230: A Key Legal Shield For Facebook, Google Is about to Change," NPR, March 21, 2018, https://www.npr.org/sections/alltech considered/2018/03/21/591622450/section-230-a-key-legal-shield-for-facebook -google-is-about-to-change.

14. Philip Napolin and Royn Caplan, "Why Media Companies Insist They're Not Media Companies, Why They're Wrong, and Why It Matters," *First Monday* 22, no. 5 (May 1, 2017), https://firstmonday.org/ojs/index.php/fm/article/view /7051/6124#p2.

15. Jason Koebler and Joseph Cox, "The Impossible Job: Inside Facebook's Strug- gle to Moderate Two Billion People," *Vice*, August 23, 2018, https://motherboard .vice.com/en_us/article/xwk9zd/how-facebook-content-moderation-works.

16. Shannon Liao, "Mark Zuckerberg Calls Tim Cook's Comments on Facebook 'Extremely Glib,'" *The Verge*, April 2, 2018, https://www.theverge.com/2018/4/2 /17188660/mark-zuckerberg-tim-cook-comments-facebook-extremely-glib.

17. Julia Carrie Wong, "Apple's Tim Cook Rebukes Zuckerberg over Facebook's Business Model," *Guardian*, March 28, 2018, https://www.theguardian.com /technology/2018/mar/28/facebook-apple-tim-cook-zuckerberg-business-model.

18. Mark Zuckerberg, "Mark Zuckerberg: The Internet Needs New Rules. Let's Start in These Four Areas," *Washington Post*, March 30, 2019, https://www.washington post.com/opinions/mark-zuckerberg-the-internet-needs-new-rules-lets-start-in -these-four-areas/2019/03/29/9e6f0504-521a-11e9-a3f7-78b7525a8d5f_story. html?utm_term=.240d1fdb9b06.

19. Mike Isaac, "Mark Zuckerberg's Call to Regulate Facebook, Explained," *New York Times*, March 30, 2019, https://www.nytimes.com/2019/03/30/technology /mark-zuckerberg-facebook-regulation-explained.html.

20. Ben Brody, "Zuckerberg's Calls for Regulation Are Seen Missing the Mark," *Bloomberg*, April 1, 2019, https://www.bloomberg.com/news/articles/2019-04 -01/zuckerberg-s-calls-for-regulation-are-seen-missing-the-mark.

21. Nick Davies and Amelia Hill, "Missing Milly Dowler's Voicemail Was Hacked by News of the World," *Guardian*, July 4, 2011, https://www.theguardian .com/uk/2011/jul/04/milly-dowler-voicemail-hacked-news-of-world.

22. Erin McCormick and N. Craig Smith, "Volkwagen's Emissions Scandal: How Could It Happen?," INSEAD, 2018. Case.

23. Luann J. Lynch and Cameron Cutro, "The Wells Fargo Banking Scandal," Darden School, 2017. Case.

24. UK Phone Hacking Scandal Fast Facts by CNN Library, CNN, April 29, 2019.

25. McCormick and Smith, "Volkswagen's Emissions Scandal."

26. How Wells Fargo's Cutthroat Corporate Culture Allegedly Drove Bankers to Fraud by Bethany Mclean, Vanity Fair May 31, 2017 https://www.vanityfair.com /news/2017/05/wells-fargo-corporate-culture-fraud

27. Sharon Waxman, "In Testimony, It's Rupert Murdoch the CEO Dilettante," *Reuters*, July 19, 2011, https://www.reuters.com/article/idUS23490391982011 0719.

28. Brian Tayan, "The Wells Fargo Cross-Selling Scandal," *Stanford Closer Look Series*, January 8, 2019.

29. Renae Merle, "After Years of Apologies for Customer Abuses, Wells Fargo CEO Tim Sloan Suddenly Steps Down," *Washington Post*, March 28, 2019.

30. McCormick and Smith, "Volkswagen's Emissions Scandal."

31. Richard Bohmer, Amy Edmondson, and Michael A. Roberto, "Columbia's Final Mission," Harvard Business School, May 4, 2010. Case.

32. Bohmer, Edmondson, and Roberto, "Columbia's Final Mission."

33. Steve Hall, "Jean-Francois Baril's Five Truths of Procurement Leadership," *Procurement Leaders*, March 20, 2012.

34. Mara Hvistendahl, "Inside China's Vast New Experiment in Social Ranking," *Wired*, December 14, 2017, https://www.wired.com/story/age-of-social-credit.

35. McKenzie Rees, Ann Tenbrunsel, and Max Bazerman, "Bounded Ethicality and Ethical Fading in Negotiations: Understanding Unintended Unethical Behavior," *Academy of Management Perspectives* 33, no. 1 (February 28 2019): 1–17.

36. Ann Tenbrunsel, Kristina Diekmann, Kimberly Wade-Benzoni, and Max Bazerman, "The Ethical Mirage: A Temporal Explanation as to Why We Are Not as Ethical as We Think We Are," *Research in Organizational Behavior* 30 (2010): 153–173.

37. Rob Tornoe, "What Happened at Starbucks in Philadelphia?," *The Inquirer*, April 16, 2018, https://www.philly.com/philly/news/starbucks-philadelphia-arrests-black-men-video-viral-protests-background-20180416.html.

38. Matthew Dollinger, "Starbucks, 'the Third Place,' and Creating the Ultimate Customer Experience," *Fast Company*, June 11, 2008, https://www.fastcompany.com/887990/starbucks-third-place-and-creating-ultimate-customer-experience.

39. Rod Wagner, "The Philadelphia Incident Was Terrible; Starbucks' Response Was Admirable," *Forbes*, June 1, 2018, https://www.forbes.com/sites/roddwagner/2018/06/01/the-philadelphia-incident-was-terrible-starbucks-response-was-admirable/#4cd5a45c23bc.

40. Nick Vadala, "Stephen Colbert Slams Starbucks over Philly Arrests on 'Late Show,'" *The Inquirer*, April 20, 2018, https://www.philly.com/philly/entertainment/celebrities/starbucks-stephen-colbert-arrests-training-20180420.html.

41. Sam Sanders, "Starbucks Will Stop Putting the Words 'Race Together' on Cups," *NPR*, March 22, 2015, https://www.npr.org/sections/thetwo-way/2015/03/22/394710277/starbucks-will-stop-writing-race-together-on-coffee-cups.

42. Rees et al., "Bounded Ethicality and Ethical Fading in Negotiations."

43. Varda Liberman, Steven M. Samuels, and Lee Ross, "The Name of the Game: Predictive Power of Reputation versus Situational Labels in Determining Prisoner's Dilemma Game Moves," *Personality and Social Psychology Bulletin* 30, no. 9 (September 1, 2004): 1175–1185.

44. Kimberly Wade-Benzoni, Ann E. Tenbrunsel, and Max H. Bazerman, "Egocentric Interpretations of Fairness in Asymmetric, Environmental Social

Dilemmas: Explaining Harvesting Behavior and the Role of Communication," *Organization Behavior and Human Decision Processes* 67, no. 2 (1996): 111–126.

45. David Lehman, Kieran O'Connor, Balázs Kovács, and George Newman, "Authenticity," *Academy of Management Annals* 13, no. 1 (2019): 1–42.

46. Lehman et al., "Authenticity."

47. Ariana Brockington, "Apple's Tim Cook Slams Facebook: Privacy 'Is a Human Right,' 'A Civil Liberty,'" *Variety*, March 28, 2018.

48. Ellen Nakashima, "Apple Vows to Resist FBI Demand to Crack iPhone Linked to San Bernardino Attacks," *Washington Post*, February 17, 2016, https://www .washingtonpost.com/world/national-security/us-wants-apple-to-help-unlock -iphone-used-by-san-bernardino-shooter/2016/02/16/69b903ee-d4d9-11e5 -9823-02b905009f99_story.html?utm_term=.52b69d695d16.

49. Mike Isaac, "Apple Shows Facebook Who Has the Power in an App Dispute," *New York Times*, January 31, 2019, https://www.nytimes.com/2019/01/31 /technology/apple-blocks-facebook.html?action=click&module=News&pgtype =Homepage.

50. Apple, *Apple's 2019 Definitive Proxy Statement Pursuant to Section 14(a) of the Securities Exchange Act of 1934*, 2019, https://www.sec.gov/Archives/edgar/data /320193/000119312517380130/d400278ddef14a.htm.

51. "Mission 2016: The Future of Strategic Natural Resources Project at MIT," 2016, http://web.mit.edu/12.000/www/m2016/finalwebsite/problems/ree.html.

52. Liberman et al., "The Name of the Game."

53. Tenbrunsel et al., "The Ethical Mirage."

54. Ting Zhang, Francesca Gino, and Joshua Margolis, "Does 'Could' Lead to Good? On the Road to Moral Insight," *Academy of Management Journal* 61, no. 3 (2018): 857–895.

Chapter 10

1. James Manyika, Susan Lund, Michael Chui, Jacques Bughin, Jonathan Woetzel, Parul Batra, Ryan Ko, and Saurabh Sanghvi, "Jobs Lost, Jobs Gained: Workforce Transitions in a Time of Automation," *McKinsey Global Institute*, December 2017, https://www.mckinsey.com/~/media/mckinsey/featured%20 insights/Future%20of%20Organizations/What%20the%20future%20of%20 work%20will%20mean%20for%20jobs%20skills%20and%20wages/MGI-Jobs -Lost-Jobs-Gained-Report-December-6-2017.ashx.

2. Paul Davidson, "More High Schools Teach Manufacturing Skills," *USA Today*, November 12, 2014, https://www.usatoday.com/story/money/business /2014/11/12/high-schools-teach-manufacturing-skills/17805483.

3. Nicholas Wyman, "Why We Desperately Need to Bring Back Vocational Training in Schools," *Forbes*, September 1, 2015, https://www.forbes.com/sites /nicholaswyman/2015/09/01/why-we-desperately-need-to-bring-back-voca tional-training-in-schools/#7e9e986887ad.

4. Clifford Krauss, "Texas Oil Fields Rebound from Price Lull, but Jobs Are Left Behind," *New York Times*, February 19, 2017, https://www.nytimes.com/2017 /02/19/business/energy-environment/oil-jobs-technology.html.

5. Ramchandran Jaikumar, "From Filing and Fitting to Flexible Manufacturing: A Study in the Evolution of Process Control," Foundations and Trends(R) in Technology, Information and Operations Management 1, no. 1 (2005): 1–120, https://ideas.repec.org/a/now/fnttom/0200000001.html.

6. Manyika et al., "Jobs Lost, Jobs Gained."

7. Dave Eggers, *The Circle* (New York: Vintage Books, 2014).

8. Mike Juang, "A New Kind of Auto Insurance Technology Can Lead to Lower Premiums, but It Tracks Your Every Move," *CNBC*, October 5, 2018, https:// www.cnbc.com/2018/10/05/new-kind-of-auto-insurance-can-be-cheaper-but -tracks-your-every-move.html.

9. Sapna Maheshwari, "How Smart TVs in Millions of US Homes Track More Than What's on Tonight," *New York Times*, July 5, 2018, https://www.nytimes .com/2018/07/05/business/media/tv-viewer-tracking.html.

10. Bidhan Parmar and Edward Freeman, "Ethics and the Algorithm," *Sloan Management Review*, Fall 2016.

11. George Westerman, "Why Digital Transformation Needs a Heart," *Sloan Management Review*, Fall 2016.

12. Greg Nichols, "Workers Don't Fear Automation (Because They Don't Understand It)," *ZDNet*, December 6, 2017, https://www.zdnet.com/article/workers -dont-fear-automation-because-they-dont-understand-it.

13. Brent Clark, Christopher Robert, and Stephen Hampton, "The Technology Effect: How Perceptions of Technology Drive Excessive Optimism," *Journal of Business and Psychology* 31, no. 1 (2016): 87–102.

14. Kimberly D. Elsbach and Ileana Stigliani, "New Information Technology and Implicit Bias," *Academy of Management Perspectives* 33, no. 2 (May 1, 2019): 185–206.

15. Robert Lowe and Arvids Ziedonis, "Overoptimism and the Performance of Entrepreneurial Firms," *Management Science* 52, no. 2 (2006): 173–186.

16. Vilhelm Carlström, "This Finnish Company Just Made an AI Part of the Management Team," *Business Insider Nordic*, October 17, 2016, https://nordic .businessinsider.com/this-finnish-company-just-made-an-ai-part-of-the-man agement-team-2016-10.

17. Roberto, *Unlocking Creativity*.

18. Tobias Baer and Vishnu Kamalnath, "Controlling Machine-Learning Algorithms and Their Biases," *McKinsey Quarterly*, November 2017, https://www .mckinsey.com/business-functions/risk/our-insights/controlling-machine -learning-algorithms-and-their-biases.

19. Maria Korolov, "AI's Biggest Risk Factor: Data Gone Wrong," *CIO Magazine*, February 13, 2018, https://www.cio.com/article/3254693/artificial-intelligence /ais-biggest-risk-factor-data-gone-wrong.html.

20. Christopher Heine, "Microsoft's Chatbot 'Tay' Just Went on a Racist, Misogynistic, Anti-Semitic Tirade," *AdWeek*, March 24, 2016, https://www.adweek.com /digital/microsofts-chatbot-tay-just-went-racist-misogynistic-anti-semitic-tirade -170400.

21. Natasha Singer, "Amazon's Facial Recognition Wrongly Identifies 28 Lawmakers, A.C.L.U. Says," *New York Times*, July 26, 2018, https://www.nytimes.com /2018/07/26/technology/amazon-aclu-facial-recognition-congress.html.

22. Joy Buolamwini, "How I'm Fighting Bias in Algorithms," TEDxBeacon Street, https://www.ted.com/speakers/joy_buolamwini.

23. "For Artificial Intelligence to Thrive, It Must Explain Itself," *Economist*, February 15, 2018, https://www.economist.com/science-and-technology/2018/02 /15/for-artificial-intelligence-to-thrive-it-must-explain-itself.

24. Haslina Ali and Rubén Mancha, "Coming to Grips with Dangerous Algorithms," *Sloan Management Review*, Fall 2016.

25. Joy Buolamwini, "How I'm Fighting Bias in Algorithms," TED Talk, updated March 29, 2017, https://www.youtube.com/watch?v=UG_X_7g63rY.

26. Parmar and Freeman, "Ethics and the Algorithm."

27. Reid Hoffman, "Using Artificial Intelligence to Set Information Free," *Sloan Management Review*, Fall 2016.

28. OpenAI, "Better Language Models and Their Implications," February 14, 2019, https://openai.com/blog/better-language-models/#sample8.

29. Rachel Metz, "This AI Is So Good at Writing That Its Creators Won't Let You Use It," *CNN Business*, February 18, 2019, https://www.cnn.com/2019/02/18 /tech/dangerous-ai-text-generator/index.html.

30. Gina Kolata, Sui-Lee Wee, and Pam Belluck, "Chinese Scientist Claims to Use Crispr to Make First Genetically Edited Babies," *New York Times*, November 26, 2018, https://www.nytimes.com/2018/11/26/health/gene-editing-babies -china.html.

31. "Can We Trust Ourselves When It Comes to Gene Editing?," Bayer AG, November 15, 2018, https://www.canwelivebetter.bayer.com/innovation/can-we -trust-ourselves-when-it-comes-gene-editing?ds_rl=1259492&gclid=EAIaIQobCh MIyuWWuZrQ4AIViq_ICh1DKAYKEAMYASAAEgI16vD_BwE&gclsrc=aw.ds.

32. Arthur C. Clarke, "Hazards of Prophecy: The Failure of Imagination," in *Profiles of the Future: An Enquiry into the Limits of the Possible* (New York: Harper and Row, 1973 [originally published in 1962]).

33. Abby Abazorius, "How Data Can Change the World," *MIT News*, September 26, 2016, http://news.mit.edu/2016/IDSS-celebration-big-data-change-world-0926.

34. Ian Chipman, "How Data Analytics Is Going to Transform All Industries," *Stanford Engineering Research & Ideas*, February 23, 2016, https://engineering .stanford.edu/magazine/article/how-data-analytics-going-transform-all-industries.

35. "Big Data: How Data Analytics Is Transforming the World," The Great Courses, 2014, https://guidebookstgc.snagfilms.com/1382_DataAnalytics.pdf.

36. "How Is Big Data Going to Change the World?," World Economic Forum, updated December 1, 2015, https://www.weforum.org/agenda/2015/12/how-is -big-data-going-to-change-the-world.

37. Nicolaus Henke, Jacques Bughin, Michael Chui, James Manyika, Tamim Saleh, Bill Wiseman, and Guru Sethupathy, "The Age of Analytics: Competing in a Data-Driven World," December 2016, https://www.mckinsey.com/~/media /McKinsey/Business%20Functions/McKinsey%20Analytics/Our%20Insights /The%20age%20of%20analytics%20Competing%20in%20a%20data%20 driven%20world/MGI-The-Age-of-Analytics-Full-report.ashx.

38. Herman Heyns and Chris Mazzei, "Becoming an Analytics-Driven Organization to Create Value," 2015, https://www.ey.com/Publication/vwLUAssets /EY-global-becoming-an-analytics-driven-organization/%24FILE/ey-global -becoming-an-analytics-driven-organization.pdf.

39. Maria Korolov, "AI's Biggest Risk Factor."

40. Robert Austin, "Unleashing Creativity with Digital Technology," *Sloan Management Review*, Fall 2016.

41. Arthur Jago, "Algorithms and Authenticity," *Academy of Management Discoveries* 5, no. 1 (March 26, 2019): 38–56.

42. Kate Darling, "'Who's Johnny?' Anthropomorphic Framing in Human-Robot Interaction, Integration, and Policy," in *Robot Ethics 2.0*, ed. P. Lin, G. Bekey, K. Abney, and R. Jenkins (New York: Oxford University Press, 2017).

43. Kate Darling, "Why We Have an Emotional Connection to Robots," TED Salon Talks, updated September 2018, https://www.ted.com/talks/kate_darling _why_we_have_an_emotional_connection_to_robots#t-699288.

44. Leila Takayama, "What's It Like to Be a Robot?," TEDx Palo Alto, updated April 2017, https://www.ted.com/talks/leila_takayama_what_s_it_like_to_be_a_robot.

45. "Driver Deactivation Policy," Uber, updated May 29, 2019, https://help .uber.com/partners/article/driver-deactivation-policy?nodeId=ada3b961-e3c2 -48e6-ac3f-2db5936e37a9.

46. Samantha Allen, "The Mysterious Way Uber Bans Drivers," *The Daily Beast*, January 27, 2015, https://www.thedailybeast.com/the-mysterious-way-uber -bans-drivers.

47. John Koetsier, "Uber Might Be the First AI-First Company, Which Is Why They 'Don't Even Think about It Anymore,'" *Forbes*, August 22, 2018, https://www .forbes.com/sites/johnkoetsier/2018/08/22/uber-might-be-the-first-ai-first-comp any-which-is-why-they-dont-even-think-about-it-anymore/#5ca511e35b62.

48. Colin Lecher, "How Amazon Automatically Tracks and Fires Warehouse Workers for 'Productivity,'" *The Verge*, April 25, 2019, https://www.theverge.com /2019/4/25/18516004/amazon-warehouse-fulfillment-centers-productivity-firing -terminations.

49. Jago, "Algorithms and Authenticity."

50. Jürgen Brandstetter, Péter Rácz, Clay Beckner, Eduardo B. Sandoval, Jennifer Hay, and Christoph Bartneck, "A Peer Pressure Experiment: Recreation of the Asch Conformity Experiment with Robots," *2014 IEEE/RSJ International Conference on Intelligent Robots and Systems*, September 2014.

51. Kate Darling, "Extending Legal Protection to Social Robots," *IEEE Spectrum*, September 10, 2012, https://spectrum.ieee.org/automaton/robotics/artificial -intelligence/extending-legal-protection-to-social-robots.

52. Elsbach and Stigliani, "New Information Technology and Implicit Bias."

53. Amy Edmondson, "Strategies for Learning from Failure," *Harvard Business Review*, April 2011.

54. Mark Meckler and Kim Boal, "Decision Errors, Organizational Iatrogenesis and Error of the 7th Kind," *Academy of Management Perspectives*, published online October 15, 2018; in press.

55. Hope Reese, "Why Microsoft's 'Tay' AI Bot Went Wrong," *TechRepublic*, March 24, 2016, https://www.techrepublic.com/article/why-microsofts-tay-ai-bot -went-wrong.

56. Design Council, "Design Methods for Developing Services," https://www .designcouncil.org.uk/resources/guide/design-methods-developing-services.

57. Peter Bright, "Tay, the Neo-Nazi Millennial Chatbot, Gets Autopsied," *Ars Technica*, March 25, 2016, https://arstechnica.com/information-technol ogy/2016/03/tay-the-neo-nazi-millennial-chatbot-gets-autopsied.

58. Robert Waterman and Tom Peters, *In Search of Excellence* (New York: Harper & Row, 1982).

59. Caroline Nyce, "The Winter Getaway That Turned the Software World Upside Down," *The Atlantic*, December 8, 2017, https://www.theatlantic .com/technology/archive/2017/12/agile-manifesto-a-history/547715.

60. Martin Fowler, "Writing the Agile Manifesto," July 9, 2006, https://martin fowler.com/articles/agileStory.html.

61. Dominic Gates, "Flawed Analysis, Failed Oversight: How Boeing, FAA Certified the Suspect 737 MAX Flight Control System," *Seattle Times*, March 17, 2019, https:// www.seattletimes.com/business/boeing-aerospace/failed-certification-faa -missed-safety-issues-in-the-737-max-system-implicated-in-the-lion-air-crash.

62. Matt Stieb, "Report: Self-Regulation of Boeing 737 MAX May Have Led to Major Flaws in Flight Control System," *New York Magazine Intelligencer*, March 17, 2019, https://nymag.com/intelligencer/2019/03/report-the-regulatory-failures-of -the-boeing-737-max.html.

63. Andrew Tangel, Andy Pasztor and Mark Maremont, "The Four-Second Catastrophe: How Boeing Doomed the 737 MAX," *Wall Street Journal*, August 16, 2019.

64. Tangel, Pasztor, and Maremont, "The Four-Second Catastrophe."

65. Gates, "Flawed Analysis, Failed Oversight."

66. Tangel, Pasztor, and Maremont, "The Four-Second Catastrophe."

67. Tangel, Pasztor, and Maremont, "The Four-Second Catastrophe."

68. Tangel, Pasztor, and Maremont, "The Four-Second Catastrophe."

69. Amit Mukherjee, "The Case against Agility," *Sloan Management Review*, September 26, 2017.

70. Bohmer, Edmondson, and Roberto, "Columbia's Final Mission." Case.

71. Daniel Weisfield, "Peter Thiel at Yale: We Wanted Flying Cars, but We Got 140 Characters," Yale School of Management, April 27, 2013, https://som.yale.edu /blog/peter-thiel-at-yale-we-wanted-flying-cars-instead-we-got-140-characters.

Chapter 11

1. Paul Mozur, Jonah M. Kessel, and Melissa Chan, "Made in China, Exported to the World: The Surveillance State," *New York Times*, April 24, 2019, https://www .nytimes.com/2019/04/24/technology/ecuador-surveillance-cameras-police -government.html?action=click&module=Top%20Stories&pgtype=Homepage.

2. Joseph Stiglitz, "Progressive Capitalism Is Not an Oxymoron," *New York Times*, April 19, 2019, https://www.nytimes.com/2019/04/19/opinion/sunday/progres sive-capitalism.html?searchResultPosition=2.

Index

Academic bias, 64–65
Airbnb, 19
Airbus, 27
Algorithms, 143–145, 148
Amazon, 148
Ambiguity, 36, 38
American system, 2, 17
"America's Toughest Bosses," 3–4
Android, 30
Apple, 86, 132–134
App stores, 30
Artificial intelligence (AI), 24, 137–138
  and bias, 144
  bots, 28, 148
  development and deployment, 138
  job creation, 139
  OpenAI, 145
  and VUCA conditions, 147
Art of Japanese Management, The, 9
Asch, Solomon, 112
Asian executives, 44, 60–61
Assembly lines, 3
Authenticity, 131–133
Authenticity, Assumption of, 147–149
Authoritarian leadership, 3–4
Automation, 24
Automobile industry, 7–8, 27
Availability heuristic, 79

Banking values, 128
Baril, Jean-François, 50, 87, 89–90, 92–93, 127

Barsoux, Jean-Louis, 60
Bayer, 146
Bechky, Beth, 113
Behavioral diversity, 66–67
Benevolence, Assumption of, 142
Benioff, Marc, 61–62, 119
Bennis, Warren, 77
Beretta, 1
Berlin Wall, 33
Bias
  academic, 64–65
  and artificial intelligence (AI), 144
  implicit, 71–72
  social, 45, 71
Bias toward action, 134, 152
BlackBerry, 30
Blockchain technology, 87
Blue ocean strategy, 19
Boehringer Ingelheim, 22
Boeing 737 MAX, 152–154
Boeing 787, 26–27, 86
Bots, 28, 148
Brain-computer interface
  technologies, 28
Brain trust, 103, 109
Breadth of learning, 79–80
Brown, Tim, 75–76
Built to Last, 10
Buolamwini, Joy, 144
Business education, 42, 51–52, 79,
  101–102
Business environment, 46–47

Case studies, 101–102
Catmull, Ed, 107–109, 111, 114
Cerebral work, 27–28, 31, 45, 51
Christensen, Clayton, 18–19
Circle, The, 141
Clark, Kim, 8
Clarke, Arthur C., 146
Code-share seating, 27
Coding, 64–65, 117
Cognitive trust, 92
Cold System, 94
Cold War, 33
Collaboration, 13, 48–51, 85–97
    levels of, 85–86
    and network information flow, 94–97
    partnership norms, 92
    and psychological safety, 94
    and role identification, 93–94
    as strategic choice, 88–91
    and trust, 87–88
    win-win partners, 91–92
Columbia space shuttle, 127
Command-and-control leadership, 91
Company culture, 42
Complexity, 35–36, 38
Computer-aided design (CAD) tools, 25
Conductors, orchestra, 104–105
Conformity, 112
Consistency, 72–73
Continuous improvement, 101
Controllability, Assumption of, 145–146
Convergent thinking, 150–151
Cook, Tim, 125, 132–134
Corporate boundaries, 42–43
Corporate crises, 123
"Could" mode, 134–135
Crampton, Catherine, 82
Creative Experience, 102
Creativity, 13, 53
    and change, 165
    conditions for, 113–119
    defined, 100–101
    and Design Thinking, 101

elements of, 103–109
    and leadership, 102, 109–113
    and learning, 51–52
    mindset, 109–113, 157
    and role models, 114
    space and time for, 116
Creativity gap, 53
CRISPR, 146
Cross-functional integration, 8
Crucibles of leadership, 77
Cuban missile crisis, 85
Cultural adaptation, 82–83
Cultural diversity, 60, 62

Daimler, 61
Darling, Kate, 148
Data analytics, 147
Data management, 29–30, 141
Data misappropriation, 124–125
Deal, Duane, 127
Decision making
    and assumptions, 80
    and cultural diversity, 62
    and empathy, 112
    and expertise, 55–56
    and foreign languages, 79
    by robots, 148–149
    and speed of change, 141
    standards of, 68
    and workers' groups, 7
Deep Dive, 103
Deming, W. Edwards, 6
Dempsey, Martin, 47
Design Continuum, 114
Design Thinking, 100–101, 111
De-skilling, 24, 31, 140, 156
Digital epoch, 11, 140–141
Digital technologies, 12–14, 82
    assumptions about, 142–149
    and disruption, 17–19
    implementation, 54–55
    and leadership failure, 127
    and organizational design, 46

principles of, 23–31, 33, 41
and VUCA conditions, 38–39
Directing Leaders, 103–106
Directors, film, 106
Discrimination, 45
Disruptive technologies, 17–19
Distributed leadership, 82–83
Distributed work, 26–27, 31, 42–43, 59,
  133
Divergent thinking, 150–151
Diversity, 71
  behavioral, 66–67
  cultural, 60, 62
  functional, 84
Drones, 25–26
Dry science, 25

Edmondson, Amy, 51, 93–94
Education, business, 42, 51–52, 79,
  101–102
Eggers, David, 141
Electricity grid collapse, 35, 38
Electronic product codes, 23
Emergent needs, 29–31
Empathy, 109–112
Empowering leaders, 10
English system, 2, 17
Environment, business, 46–47
Epitropaki, Olga, 102–103, 106–107
Epochal transitions, 1–2, 9
Errors, types of, 149–155
Essence of Decision, The, 85
Ethical fading, 89–90, 128–129, 131,
  133
Ethics, 144–145
Ethnorelativity, 67
Evalueserve, 76
Executives. *See also* Leadership
  Asian, 44, 60–61
  and collaboration, 86
  country of origin, 44–45
  I-shaped/T-shaped, 75–76
  promotion rates, 60

T-prime, 76–79
  work week, 116
Exploitation, 51

Facebook, 31, 124–125, 132
Facilitating Leaders, 103, 106
Failures, 149
Fairness, 90–91, 116
Feedback, ethical, 131
Film directors, 106
Financial services industry, 36–37
Fintech, 18–19
First principles, 23
Fisher, Allan, 65
Follett, Mary Parker, 102
Ford, Henry, II, 5
Freeman, Edward, 144
FU Fund, 131
Fujimoto, Takahiro, 8
Functional diversity, 84

Galaxy Note 7, 147
Garvin, David, 7
Gelfand, Michele, 66
Gender diversity, 66
Gene-editing technology, 146
Geneen, Harold, 5
Genetic testing, 24
Genomic drugs, 20–21, 29
Gerstner, Lou, 59, 67
GlaxoSmithKline (GSK), 22
Globally integrated enterprise (GIE), 59
Global Survey, 10–12
Global work distribution, 26–27, 31
Goleman, Daniel, 4–5
Google, 64, 132
Google Duplex, 28
Gorilla Glass, 86
GPS systems, 24–25
Groysberg, Boris, 78

Hailey, Arthur, 7
Hargaden, Andrew, 113

Hinds, Pamela, 82
Hiring, 116–117
Hoegl, Martin, 10
Hoffman, Reid, 145
Hot System, 93–94
Human Genome Project, 20–21
Human needs, 54–55

IBM, 59–60
IBM 360, 9
Icelandic banks, 37
Identity, personal, 63–64
IDEO, 75–76
Implementation, 54–55
Implicit Association Test, 72
Implicit bias, 71–72
Inclusionary leadership, 13, 62–63,
    72–73, 128
Incomplete knowledge, 69–71
India, 43, 65
Infallibility, Assumption of, 143–145
Innovation, 101, 103
In Search of Excellence, 152
Integrating Leaders, 106
Integration and Values Teams (IVTs),
    59–60
Intel, 86
Intellectual property trading, 28, 85–86
Intellectual work, 27–28, 31, 45, 51
Interdependence, intensive/reciprocal, 86
Internet of Things, 30, 38, 147
Invention, 101
iOS, 30
I-shaped executives, 75

Jaikumar, Ramchandran, 1, 9, 56, 140
Japan, 6–7
    cross-functional integration, 8
    and foreign executives, 44–45
    inclusionary leadership, 63
Jobs, Steve, 86
Johnson & Johnson, 21–22, 61, 67–68
Juran, Joseph, 6

Kahneman, Daniel, 78–80
Kark, Ronit, 102–103, 106–107
Kelly, Kevin, 78
Kirikova, Vera, 60
Knowledge bases, 81
Knowledge obsolescence, 82, 105
Korhonen, Pertti, 127

Language of leadership, 68
Leadership. See also Executives
    authoritarian, 3–4
    command-and-control, 91
    and creativity, 102, 109–113
    crucibles of, 77
    current assumptions, 42–45
    distributed, 82–83
    diversity, 60
    and environment, 46
    errors of, 124
    inclusionary, 13, 62–63, 72–73, 128
    language of, 68
    national/corporate boundaries, 43–44
    types of, 103–109
    values-driven, 131–135
Learning
    breadth of, 79–80
    creative, 51–52
Levinson, Harry, 6
LGBTQ rights, 134
Lindahl, Goran, 69
Linear thinking, 80–81
Lockheed, 9
Lucas, George, 108

MacOS, 86
Mainemelis, Charalampos, 102–103,
    106–107
Manning, Christopher, 145
Manwani, Harish, 60
Margolis, Jane, 64–65, 117
MCAS system, 153
McChrystal, Stanley, 48
McDonald, Bryan, 78

Medical diagnoses, 24
Medical profession, 111–112
Mentors, 131
Microsoft Tay, 150–151
Microsoft Windows, 86
Mindset change, 157
Minimum viable product, 152
Moral hazard, 38
Multilingual capability, 77–78
Murdoch, Rupert, 126

Network diagrams, 94–97
Networked companies, 26
Network information flow, 94–97
Neural machine translation
    technology, 25
Nie, Winter, 60
Nokia, 30, 50–51, 127
Novartis, 22
Nuclear energy, 36

Omniscience, Assumption of, 146–147
OpenAI, 145
Operating systems, 86
Orchestra conductors, 104–105
Organizations
  collaboration between, 85
  conditions for creativity, 113–119
  pharmaceuticals industry, 21–23
  restructuring, 8–9
  and values, 127
  and win-win partnerships, 91–92

Pacific Asian executives, 44
Page, Scott, 71
Palmisano, Sam, 59, 67, 69, 72
Parmar, Bidhan, 144
Parsons, Richard, 90
Performance evaluation, 117–119
Personal leadership philosophy, 161
Pharmaceuticals industry, 20–23,
    26–27, 30
Physical work, 27, 31

Pichai, Sundar, 64
Piech, Ferdinand, 126
Pixar, 107–108
Prioritizing Exercise, 161–164
Prisoner's dilemma, 88–89
Privacy, 125
Procedural justice, 90–91, 116
Procedural trust, 92
Producers, 107
Producing Leaders, 107–109, 162–163
Product development, 8
Productivity mindset, 157
Psychological safety, 67, 94, 115
Purposive trust, 92

Quality Control Circles, 7
Quality movement, 6–8, 11, 17

#RaceTogether initiative, 130
Radical transparency, 30–31, 42, 45, 124
Radio-frequency identification (RFID),
    22–23
Retraining, 140
Reverse mentors, 82
Revson, Charles, 5
Rio Tinto, 60
Risk standards, 155
Robotics, 28
Robots, 147–149
Role identification, 93–94
Role modeling, 114
Ryan, James, 113

Samsung, 147
Sandberg, Sheryl, 124
Schultz, Howard, 130
Science, dry/wet, 25
Scientific management, 2–3, 11, 17
"Should" mode, 134
Simulations, 80
Skills, 24–26, 31
Skunk Works, 9
Sloan, Timothy, 126

Smartphones, 25, 30
Smith, Alvy, 108
Smith, Diana, 93
Sobbott, Susan, 83, 88
Social bias, 45, 71
Social credit systems, 87, 128
Software development, 152
Software robots, 28
Solow, Robert, 102
Specialization, 8
Stage-gate process, 109
Starbucks restroom incident, 129–130
Statistical process control, 6
Stiglitz, Joseph, 165
Stock charts, 80
Strategic intent, 83, 137, 141, 156–157
Stumpf, John, 126
Subprime mortgage crisis, 37–38
Sun Microsystems, 70
Surveyors, 25
Sustaining technologies, 18
System 1 thinking, 79

Tay chat bot, 150–151
Taylor, Frederick Winslow, 2
Teams, 9–11
  and collaboration, 86
  and trust, 92–93
Technologies. *See also* Digital
    technologies
  disruptive/sustaining, 17–19
  long-arc-of impact, 2
  quality, 7–8
Thiel, Peter, 155
Thinking gap, 52–53
Thomas, Robert, 77
3-D printing, 28
Tieto, 143
Time to expertise, 55–56
Time-and-motion studies, 3
Time Warner, 90
Tit-for-tat strategy, 89
Tolstedt, Carrie, 126

Toy Story, 107–108
Toy Story 2, 111
T-prime executives, 76–79
Transistor radios, 18
Translation technology, 25
Transparency, 30–31, 42
Trippe, Juan, 5
Trust, 87–88, 92–93
T-shaped executives, 75–76
Tversky, Amos, 78
Twitter, 150

Uber, 18–19, 148
Uncertainty, 35, 38
Uncrossable lines, 132
Unfairness, 68–69
Uniformity, 112–113
Unilever, 60–61
Unpredictable needs, 29–30
Up-skilling, 25–26, 31, 140, 156
US Army, 47–48

Validation, 115–116
Values, 14, 127–135
ValuesJam, 60
Van Wersch, Wouter, 77
Venture capital, 124
Virtual reality technology, 156
Volatility, 35, 38
Volkswagen, 126
Vollenweider, Marc, 76–78
Volvo, 9
VUCA (volatile, uncertain, complex,
    and ambiguous) conditions, 34
  and AI, 147
  and army leadership, 47–48
  and assumptions, 80
  and business environment, 46–47
  and digital technologies, 38–39
  and subprime mortgage crisis, 37–38

"Want" mode, 132, 134
Watkins, Michael, 55, 78, 161

Weapons manufacturing, 1
Weinberger, David, 47
Weiss, Matthias, 10
Wells Fargo, 126
Weta, 106
Wet science, 25
Wheels, 7
Winterkorn, Martin, 126
Win-win outcomes, 89–92
Wisdom of Teams, The, 10
Women
  academic bias, 64–65
  in computing, 64–66, 117
  implicit bias, 72
  and inclusionary policies, 62
  pay disparity, 91
  social bias, 45
Work
  cerebral, 27–28, 31, 45, 51
  distributed, 26–27, 31, 42–43, 59, 133
  physical, 27, 31

Xiao, Daphne, 60
Xiaoice bot, 151

Zuckerberg, Mark, 124–125
Zunger, Yonatan, 65